Leonardo Pires Carneiro

Development of an Electrochemical Biosensor Platform and a Suitable Low-Impedance Surface Modification Strategy

Schriften des Instituts für Mikrostrukturtechnik
am Karlsruher Institut für Technologie (KIT)
Band 27

Hrsg. Institut für Mikrostrukturtechnik

Eine Übersicht aller bisher in dieser Schriftenreihe erschienenen Bände finden Sie am Ende des Buchs.

Development of an Electrochemical Biosensor Platform and a Suitable Low-Impedance Surface Modification Strategy

by

Leonardo Pires Carneiro

KIT Scientific Publishing

Dissertation, Karlsruher Institut für Technologie (KIT)
Fakultät für Maschinenbau

Tag der mündlichen Prüfung: 19. August 2014
Hauptreferent: Prof. Dr. Volker Saile
Korreferent: Prof. Dr. Axel Rosenhahn

Impressum

KIT Scientific Publishing

Karlsruher Institut für Technologie (KIT)
KIT Scientific Publishing
Straße am Forum 2
D-76131 Karlsruhe

KIT Scientific Publishing is a registered trademark of Karlsruhe Institute of Technology. Reprint using the book cover is not allowed.

www.ksp.kit.edu

Print on Demand 2014

ISSN 1869-5183
ISBN 978-3-7315-0272-2
DOI 10.5445/KSP/1000043414

Development of an Electrochemical Biosensor Platform and a Suitable Low-Impedance Surface Modification Strategy

Zur Erlangung des akademischen Grades

Doktor der Ingenieurwissenschaften

der Fakultät für Maschinenbau

Karlsruher Institut für Technologie (KIT)

genehmigte

Dissertation

von

M. Sc. Leonardo Pires Carneiro

Tag der mündlichen Prüfung: 19.08.2014
Hauptreferent: Prof. Dr. Volker Saile
Korreferent: Prof. Dr. Axel Rosenhahn

Eigentlich weiß man nur, wenn man wenig weiß.
Mit dem Wissen wächst der Zweifel.

Johann Wolfgang von Goethe

Acknowledgement

This work was created during my activities at the Institute of Microstructure Technology at Karlsruhe Institute of Technology.
I thank Professor Dr. Volker Saile and Professor Dr. Andreas Guber for the possibility to conduct this work under their supervision. I also thank Professor Dr. Axel Rosenhahn for assuming the role of second reviewer.
My special thank you goes to my advisor Dr. Bastian E. Rapp, I couldn't have asked for a better advisor! The joy und unwavering enthusiasm he has for research were contagious and always motivating, even during the toughest times of my Ph. D. I am also very grateful to Dr. Gary Davidson for his advising in the many biological questions.
I would like to thank all my colleagues for their support, encouraging discussions and fun times we had. Doro thank you for the brilliant idea!
I also acknowledge Karlsruhe House of Young Scientists (KHYS) for the funding of my stay abroad.
I am also very grateful to my friends which supported me during this journey.
Lastly, I would like to thank my family for their love and encouragement. Especially to my mother for her unconditional support and love of science and my brother for always being there for me and making me laugh even after a frustrating day.

Abstract

In this work, a low-impedance surface modification method for electrochemical impedance spectroscopy (EIS) based biosensors was developed and tested. Initially several techniques described in literature were implemented, tested and optimized for impedimetric biosensors but due to their individual limitations a novel method based on photobleaching was developed and tested.

The gold electrodes used during this work were made completely in house by means of standard photolithography and gold etching. Printed photomasks and maskless lithography were used. The electrode manufacturing method was tested and optimized on plastic, ceramics and glass substrates. Glass substrates were found to be the best substrate material due to the superior gold adhesion (using a titanium adhesion promoting layer) and low porosity.

One of the applications of these electrodes was in the detection of bacterial biofilm growth and respiratory activity by means of impedance spectroscopy and amperometry techniques. The colonization of the electrode surface by bacteria can be directly monitored as an impedance increase. The respiratory activity of a biofilm is associated with extracellular transport of charges, i.e., current, which were collected and quantified by means of amperometry. Higher currents correlate with higher respiratory activity.

Surface modification plays the most important part in a biosensor as it is responsible for providing the functional groups to immobilize the ligands and for the suppression of nonspecific adsorption. It wasn't required for the monitoring of biofilms but it is necessary for affinity assays. Surface modifications such as self-assembled monolayer (SAM), conductive polymers and electrochemical grafting are commonly used methods described in literature which were investigated and adapted during the course of this work. Five different SAM forming molecules were tested during this work among them four alkane-based compounds with different chain lengths and one aromatic compound. Additionally, two different mixed SAMs were investigated. In the course of these experiments it was found that all these SAMs are of very high surface impedance resulting in very high detection limits of 25 µg/ml for biotinylated bovine serum albumin (BSA) in phosphate-buffered saline. Experiments carried out with conductive polymers (polypyrrole) concluded that the layer conductivity is adequate but disadvantageously high levels of nonspecific adsorption were found. The limit of detection of biotin using this method was 10 ng/ml. Electrochemical grafting was examined next showing low surface impedance and moderate degrees of nonspecific adsorption but due to the lack of reproducibility and because of the complex, delicate and stringent temperature controlled deposition process it was found unsuitable for this application.

A novel surface modification method by means of fluorescein photobleaching was tested and optimized during this work. Fluorescence microscopy experiments confirmed the formation of a layer which promotes the specific binding of fluorescence marked streptavidin and a significant reduction in nonspecific adsorption. These findings were also confirmed during EIS measurements and a specific binding assay was successfully carried with a concentration of 100 pg/ml of biotin in spiked phosphate-buffered saline background medium.

Kurzfassung

Im Rahmen dieser Arbeit wurde eine niederohmige Oberflächenmodifikation für elektrochemische Impedanz-spektroskopie (EIS) entwickelt und untersucht. Zahlreiche, in der wissenschaftlichen Literatur beschriebene, Verfahren für die Oberflächenmodifikation impedimetrischer Biosensoren wurden untersucht und optimiert. Aufgrund ihrer individuellen Nachteile wurde eine neue Methode auf der Grundlage von Photobleichen entwickelt und evaluiert.

Die in dieser Arbeit verwendeten Goldelektroden wurden eigens mittels Photolithographie und Ätzen aus Gold hergestellt. Hierfür wurden sowohl gedruckte Photomasken als auch maskenlose Lithographie verwendet. Das Herstellungsverfahren für die Goldelektroden wurde im Hinblick auf verschiedene, im Rahmen der Arbeit verwendeten Materialien wie Kunststoffe, Keramik- und Glas hin optimiert. Glassubstrate haben sich dabei aufgrund ihres höheren Adhäsionsvermögens (unter Verwendung einer Titanhaftschicht) und geringeren Porosität als bestes Material erwiesen.

Eine der Hauptanwendungsfelder dieser Arbeit war die Überwachung von Wachstum und Stoffwechsel bakterieller Biofilme mittels Impedanzspektroskopie und Amperometrie auf technischen Sub-straten. Die bakterielle Besiedlung von der Elektrodeoberfläche kann direkt über einen Anstieg der Impedanz erfasst werden. Der Stoffwechsel des Biofilms korreliert mit dem extrazellulären Ladungstransport, der als Strom mittels Amperometrie gemessen wurde. Höhere Ströme weisen auf erhöhten Stoffwechsel hin.

Oberflächenmodifikation ist von besonderer Bedeutung bei der Entwicklung von Biosensoren denn diese Beschichtungen stellen funktionelle Gruppe für die Anbindung von Liganden zur Verfügung und sorgen für die Unterdrückung von nichtspezifischer Adsorption. Diese

Eigenschaften sind vor allem für Affinitätsassays wichtig, für die Überwachung von Biofilmen ist dies keine notwendige Voraus-setzung. Self-assembled monolayer (SAM), leitfähige Polymere und elektrochemisches Grafting sind Bespiele von literaturbeschriebene Methoden für Oberflächemodifikation, die im Rahmen dieser Arbeit untersucht und angepasst wurden. Fünf verschiedene SAM-bildende Moleküle wurden während dieser Arbeit untersucht. Hierbei wurden vier alkane Mercaptosäuren mit verschiedenen Kettenlängen, sowie eine aromatische Mercaptosäure untersucht. Ergänzend wurden zwei gemischte SAMs untersucht. Im Rahmen dieser Experimente wurde festgestellt, dass alle SAMs hochohmige Schichte erzeugen und daher nur hohe Nachweisgrenzen (bestes Resultat: 25 µg/ml von biotinyliertes Rinderserumalbumin in Phosphatgepufferte Kochsalzlösung) ermög-lichen. Experimente mit leitfähigen Polymeren (Polypyrrol) haben gezeigt, dass die Leitfähigkeit der Polymerschicht ausreichend hoch war, die nichtspezifische Adsorption allerdings für Anwendungen in der Biosensorik zu hoch ist. Die Nachweisgrenze für Biotin für diese Methode betrug 10 ng/ml in Pufferlösung. Elektrochemisches Grafting erlaubte sowohl die Herstellung niederohmiger Oberflächen als auch eine moderate Unterdrückung der nichtspezifischen Adsorption. Die Methode wurde allerdings wegen der geringen Reproduzierbarkeit, des komplexen und emp-findlichen Beschichtungsverfahrens und der notwendigen strikten Temperaturkontrolle als ungeeignet für diese Anwendung eingeordnet.

Im Rahmen dieser Arbeit wurde eine neue Methode für die Oberflächenmodifikation mittels Photobleichen von Fluorescein untersucht und optimiert. Fluorescein-konnte mittels dieses Verfahrens auf vorbehandelten Elektroden immobilisiert und mittels Fluoreszenzmikroskopie detektiert werden. Diese Schicht hat eine wesentliche Unterdrückung der nichtspezifischen Adsorption bei hoher Bindungskapazität für die Liganden gezeigt. Impedanz-spektroskopie unter Verwendung von Elektroden, die mittels dieser Technik funktionalisiert wurden, erlaubten eine Detektion von Biotin mit einer Nachweisgrenze von 100 pg/ml in Puffer.

Table of contents

Table of figures

List of tables

List of abbreviations

AIDS	Acquired immunodeficiency syndrome
11-MUA	11-Mercaptoundecanoic acid
11-MUE	11-Mercaptoundecanol
3-MPA	3-Mercaptopropionic acid
3D	Three-dimensional
4-ABA	4-Aminobenzoic acid
4-MBA	4-Mercaptobenzoic acid
6-MHA	6-Mercaptohexanoic acid
8-MOA	8-Mercaptooctanoic acid
b-BSA	Biotinylated bovine serum albumin
BHI	Brain hearth infusion
BSA	Bovine serum albumin
CAD	Computer-aided design
COC	Cyclic olefin copolymer
CV	Cyclic voltammetry
CVD	Chemical vapor deposition
DC	Direct current

DMSO	Dimethyl sulfoxide
DNA	Deoxyribonucleic acid
EIS	Electrochemical impedance spectroscopy
ELISA	Enzyme-linked immunosorbent assay
EPS	Extracellular polymeric substances
FITC	Fluorescein isothiocyanate
GDP	Gross domestic product
HF	Hydrofluoric acid
HIV	Human immunodeficiency virus
MEMS	Microelectromechanical system
PBS	Phosphate-buffered saline
PDMS	Polydimethylsiloxane
PEG	Poly(ethylene glycol)
PTFE	Polytetrafluoroethylene
QCM	Quartz crystal microbalance
SAW	Surface acoustic wave
SPR	Surface plasmon resonance
TRIS	Tris(hydroxymethyl)aminomethane
TBST	Tris-Buffered Saline and Tween 20

1 Introduction

Microsystem technology has been gaining attention not only in industrial applications but also in everyday life products. The ever increasing miniaturization and integration of technical systems such as those present in a smartphone, e.g., touchscreen, radio modules, sensors, etc., are typical application areas of microsystem technology. Additionally, biomedical systems such as hearing aids, pacemakers and biosensors also constitute a very important segment of microsystem technology. The global market for biosensors is following a steep upward trend; in 2000 biosensors represented a market worth of almost 2 billion US dollars, ten years later this share rose to over 13 billion dollars and the market predictions for 2018 show figures around the 17 billion dollar mark [1]. Although a wide range of biosensors applications have been reported in literature, recent data and analysis of potential markets suggest that the key application area will remain to be healthcare, more specifically diagnostics for disease treatment and prevention [2]. Developing countries face diseases such as HIV/AIDS and diarrheal infections which kill 1.8 and 2.5 million people yearly, respectively. The overall population growth and increased longevity are causing a rash demographic change towards a rapidly aging population. Since the 1980 the population over 60 years has doubled and, according to predictions, should reach the 2 billion mark by 2050 [1]. All these effects combined are leading to an ever-increasing healthcare spending. According to the WHO [3], the USA spends almost 18 % of its Gross Domestic Product (GDP) on healthcare and the average European Union country almost 10 %.

The development of biosensors was driven by the glucose measurement system for diabetic patients during the middle of the 1970's. This marked has been expanding and diversifying ever since. Today's biosensor systems are used in the most diverse fields such as drug discovery, detection of environmental pollutants (e.g., pesticides), routine labora-laboratory assays monitoring medically relevant molecules (e.g., hormones, protein, DNA) and the detection of illegal and therapeutic drugs.

Detection principles used by biosensors are very diverse, ranging from gravimetric (e.g., quartz crystal microbalance and surface acoustic waves), calorimetric (e.g., thermal sensors), optical (e.g., fluorescence microscopy and surface plasmon resonance) as well as electrochemical methods (e.g., potentiometry, amperometry and impedance spectroscopy). The most widely known biosensor, the glucose measurement system, is based on an electrochemical method (amperometry). These methods currently hold over 85% of the global market share of biosensors [1]. Electrochemical methods have been gaining increasing attention recently due to their simplicity, high sensitivity, low-cost and suitability for miniaturization.

Detection principles can be divided in direct and indirect assay formats. Indirect detection principles require some sort of label (e.g., enzyme, radioactive, fluorescence or visible light marker) while direct approaches detect the desired analyte directly usually by means of detecting the actual binding of the analyte to a sensor surface. Indirect assays have the disadvantages of requiring a, sometimes costly, label which has to be coupled to the target analyte thus requiring an additional chemical reaction prior to the assay. This reaction increases the assay costs and the total sample-to-response time thus hindering the use of such systems in real-time sensing applications.

At the heart of almost every biosensor is the sensing element which normally consists of suitable ligands or receptors to allow specific interaction with the biological system, i.e., with the target analyte. The sensing element is coupled to a physicochemical transducer which is responsible for translating the interaction of ligand and analyte, i.e., the binding event, into a physically detectable quantity such as, e.g., a voltage or a current. The methods used to create the sensing element are termed surface modification and are very diverse. A few examples from literature are: self-assembled monolayers (SAM), conductive polymers, entrapment in a polymer matrix, chemical grafting and affinity binding using suitable functional groups. Each method has its advantages and disadvantages according to desired physicochemical properties (i.e., surface conductivity, transparency, chemical stability, etc.).

Microfluidics plays an important part in clinical diagnostics and in chemical analysis. A microfluidic system possesses the ability to handle (i.e., move, mix and separate) very small amounts of fluids. This can be advantageous for a number of applications which have to do with very limited amounts of sample or with costly reagents. Furthermore, in case biological samples are to be tested, the smaller the amount of sample (e.g., blood) required the less difficult the sample acquisition from the patient's body will be.

Electrochemical methods have been benefiting from advances in the field of microelectronics, especially from advances in fabrication techniques such as photolithography. This technique has contributed greatly to not only the miniaturization of the measurement electronics but also of the electrodes themselves. The combination of microfluidics and microelectrodes has enabled electrochemical systems to be more compact and portable resulting in, e.g., modern glucose test devices which allow new applications such as, e.g., subcutaneous continuous glucose measurement [4].

This work encompassed the development of an electrochemical biosensor platform including the electrode fabrication process, microfluidic integration and development of a suitable low-impedance surface modification strategy. Experiments based on fluorescence microscopy and electrochemical impedance spectroscopy have been used to confirm the suitability of this platform.

This work is structured in six chapters. In chapter 2 the theory of biosensors, with a focus on electrochemical biosensors, as well as the concepts of photolithography will be explained. In chapter 3 the required materials and fabrication steps for the electrode manufacturing by means of photolithography will be explained; additionally the measure-measurement setup and techniques used for the experiments will be elucidated. The different electrode surface modification protocols and the evaluation using fluorescence microscopy as well as electrochemical impedance spectroscopy will be discussed in chapter 4. Chapter 5 will give an outlook of future work. Chapter 6 summarizes this work by giving an overview and conclusion.

2 Basic principles

This chapter will elucidate the basic theoretical concepts of this work. Starting with the description, classification and history of biosensors this chapter will also explain the basic concepts of electrochemistry and surface modification, both of which were essential to the developed the biosensor used in this work. Additional concepts like biofouling, photolithography, and fluorescence microscopy are also discussed in this chapter.

2.1 Biosensors

A biosensor is an analytical device consisting of a biological sensing element and a physicochemical transducer which converts a biological signal (usually a binding event) into a physically detectable signal. A biosensor is an instrument that carries out complex bioanalytical assays (see Figure 1). The sensing element is responsible for the specificity towards the analyte at question which ranges from antibodies to enzymes, DNA fragments, proteins and cells. The transducer converts the biochemical detection event to a detectable signal. The most common transduction principles are optical, calorimetric, gravimetric, amperometric, impedimetric and potentiometric [5]. A biosensor's application spectrum comprises a wide range of tasks, ranging from clinical diagnostics, food safety, industrial processes control, pollution monitoring, drug discovery, to military and security applications [6].

The increasing interest in the fields of biosensors is reflected directly in its steep rise in publication numbers. In 1985 there were approximately 100 papers [1] on the subject and in 2011 this number rose to 4500. To put this in perspective, the papers published in 2011 alone represent more than 10% of all articles ever published about biosensors. This upward trend can also be seen in the global market for biosensors which rose from 2 billion US dollars market share in 2000 to 13 billion dollars and predictions for 2018 show figures around the 17 billion dollar mark [1]. The accelerated aging of the population is driving an increase of healthcare expenditure globally.

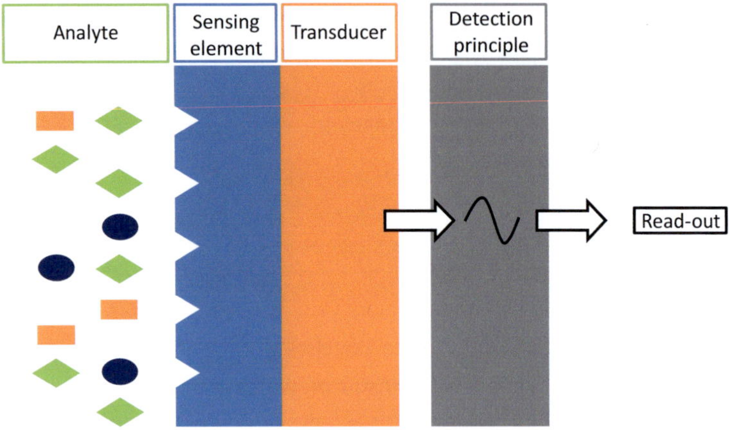

Figure 1 – Schematic diagram of working principle of a biosensor. The biosensor consists of a sensing element and a transducer. The former allows for the biochemical interaction with the analyte whilst the latter converts this interaction to an easily detectable signal.

As previously listed there are several analyte specific binding partners that are used as a sensing element in biosensors. Literature examples encompass cells [7], antibodies [8], enzymes [9] and nucleic acids [10] as common sensing elements for biosensors. These sensors can be further categorized in direct and indirect assay formats. Direct biosensors are able to directly detect the binding event. Indirect biosensors require a second binding event, usually the attachment of a suitable label, to the previously bound analyte. Only this additional label can be detected in indirect biosensors. This additional label requires a secondary reaction which increases the complexity of the assay as well as its total time and overall costs.

2.1.1 History and application of biosensors

Leyland C. Clark is commonly referred to as the father of the biosensor concept. Having worked extensively with electrochemical methods for oxygen detection he published in 1956 a landmark paper on his so called "oxygen electrode" [11]. Based on this electrode he elucidated the first biosensor concept in a talk at the New York Academy of Sciences symposium in 1962 describing how to modify his oxygen electrode by adding enzyme transducers. In the paper published in the same year, Clark [12] demonstrated how glucose oxidase was entrapped at an electrode using a dialysis membrane. Using this device, he could correlate a decrease in the measured oxygen levels with the concentration of glucose using amperometric methods. Biosensors using alternative detection principles followed. In 1969 Guilbault and Montalvo published a paper [13] on a potentiometric urea sensor based on the immobilization of urease on ammonium-selective liquid membrane electrodes. Five years later the use of thermal transducers was proposed by Cooney [14]. In 1975 Lubbers and Opitz proposed a fiber optic sensor with an immobilized indicator which could measure carbon dioxide and oxygen [15]. In the same year another technique was proposed by Diviès who proposed using bacteria for measuring alcohol concentration using what he called "microbial electrodes" [16].

In 1975 the first commercial glucose biosensor system from Yellow Springs Instruments was commercialized. But it was not until the 1980s that biosensors were produced on a larger scale. Yellow Springs Instruments had established a steady business but it was nowhere close to the anticipated success. This was mainly due to the high production costs which rendered them uncompetitive with mass testing technologies e.g., spectrophotometry, radioimmunoassay, enzyme immunoassay and enzyme-linked immunosorbent assay (ELISA) [17, 18]. In 1984 Anthony P. F. Turner presented an immobilized mediator based on ferrocene which enabled the fabrication of inexpensive enzyme electrodes [19] . This led to the development of cheap, portable screen-printed enzyme electrodes which were commercialized three years later by MediSense (Cambridge, MA) combined with a pen-blood-sized glucose monitoring device. This seminal work paved the way for a multi-billion dollar market which uses devices based on this method almost unchanged until today [20].

The idea of immobilizing antibodies to a transducer for the direct antigen detection has been explored since the early 1970s but it was a seminal paper by Liedberg et al. in 1983 [21] that paved the way for the commercial success of immunosensors. The authors described a real-time affinity binding technique using surface plasmon resonance [22]. Seven years later the BIAcore system was launched based on this detection technique. Today it has become the gold standard for direct sensing applications and is used as reference analytical technique for many new detection systems. Due to the large size of a commercially available SPR device (bench-top or floor-standing) weighing several kilos it is not normally used as a point-of-care device.

2.1.2 Electrochemical biosensors

As previously mentioned the first measurement principle used for biosensing was based on electrochemistry. This detection principle is still the most widely used principle with over 85% of the global market share of biosensors. Advantages of this method range from simplicity, low-cost and suitability for miniaturization to high sensitivity and accuracy. Electrochemical biosensors can use a multitude of transduction principles transforming a biological binding event into an evaluable electric signal. The most commonly found principles in literature are potentiometric, amperometric and impedimetric biosensors. The analytical information obtained by potentiometry is based on converting the biochemical binding event into an electrical potential, mostly by using ion-selective electrodes [23]. In amperometry a constant potential is applied to an electrode while measuring the current resulting from reduction or oxidation of an electroactive species, mostly an enzyme, immobilized on top of an electrode [24]. Impedimetry is based on applying an alternating voltage while monitoring the resulting current thus obtaining the impedance of the layer on top of the electrodes in question [25]. By coupling a biorecognition element to an electrode it is possible to monitor the change of impedance, and consequently the binding event, during an essay.

2.1.3 Electrochemical impedance spectroscopy (EIS)

EIS is a powerful measurement technique capable of analyzing the complex electrical resistance, i.e., the impedance, of a system and assessing its surface and bulk electrical properties. By applying a small amplitude alternating current (AC) voltage to an electrode pair and measuring the resulting current one can calculate the impedance as follows:

$$Z(j\omega) = \frac{V(j\omega)}{I(j\omega + \phi)} = |Z(\omega)|e^{j\phi} \quad (1)$$

Here, Z is the impedance, V the voltage, I the current, j the imaginary unit and ω the angular frequency.

Since EIS deals with complex quantities, its history began with the introduction of impedance in electrical engineering in 1880 by the British self-taught electrical engineer and mathematician Oliver Heaviside. It was only in 1899 that the German physicist, Emil G. Warburg, extended the concept of impedance to electrochemical systems by deriving the impedance of a diffusional transport of an electroactive species to an electrode surface. This process still carries his name and the circuit element used in an equivalent circuit diagram is referred to as Warbug impedance. The invention of the potentiostat 41 years later and the development of frequency response analyzers during the 1970s pushed the use of EIS towards the exploration of corrosion mechanisms [26]. Its success in this field led to an explosion in the use of EIS for a wide range of applications ranging from rheological characterization of fluids, battery and fuel-cell degradation monitoring to protein detection and immunosensing which is the main topic of this work.

As the name implies, impedance values are obtained over a frequency range which in turn generates an impedance spectrum. This spectrum allows for a characterization of the system at hand, e.g., surfaces, layers, diffusion processes, etc. Figure 2 displays the temporal representation of the current and voltage (see Figure 2 a) as well as two different methods to represent complex impedance values; one based on amplitude and phase components (see $|Z|$ and φ in Figure 2 b) whilst the other displays the data on real and imaginary axis (see Z_R and Z_I in Figure 2 b).

Those two different representations of complex impedance values led to the creation of two different representations of complex impedance spectra. One based on phase and magnitude information, called Bode plot and the other on real (resistance) and imaginary (reactance) components, termed the Nyquist plot (see Figure 3 c/d for a Bode plot and Figure 3 b for a Nyquist plot). The ohmic resistance is represented by R and C stands for an ideal capacitor. Due to the simplicity of the Nyquist plot and the fact that it requires only one diagram (Bode diagrams require separate plots for phase and magnitude over frequency), the former is overwhelmingly found in literature.

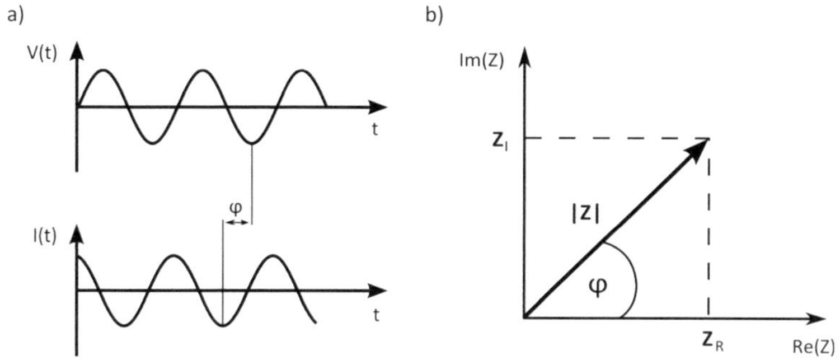

Figure 2 – Schematic diagram of ways to display complex impedance values. a) Temporal representation of the current and voltage with a phase shift φ. b) The impedance can be represented by modulus (|Z|) and phase difference between the voltage and current (φ) or by its real (Z_R) and imaginary (Z_I) parts.

Figure 3 exemplifies the differences between Nyquist (Figure 3 b) and Bode (Figure 3 c and d) diagrams for the same circuit. The circuit in question is a resistor in parallel with capacitor (Figure 3 a). The Bode diagram requires two different plots (or curves) to be interpreted. Furthermore, it lacks the distinction between real and imaginary part of the impedance. The amplitude plot seen in Figure 3 c can be easily explained. At low frequencies the combined impedance of the RC parallel branch is equal to the resistance of the resistor as the capacitor impedance tends to infinity (see Table 1 for formula). The increase in frequency causes the decrease of the capacitor impedance which tends to zero at high frequencies. The Nyquist plot (Figure 3 b) requires only one curve but omits the direct frequency information as only the direction of the frequency growth can be shown. However, it is easy to visualize that at low frequencies the impedance of this circuit tends to R whilst at low high frequencies it tends to zero. Because the exact frequency information is not required for an initial analysis of the measured dataset only Nyquist plots are used in this work.

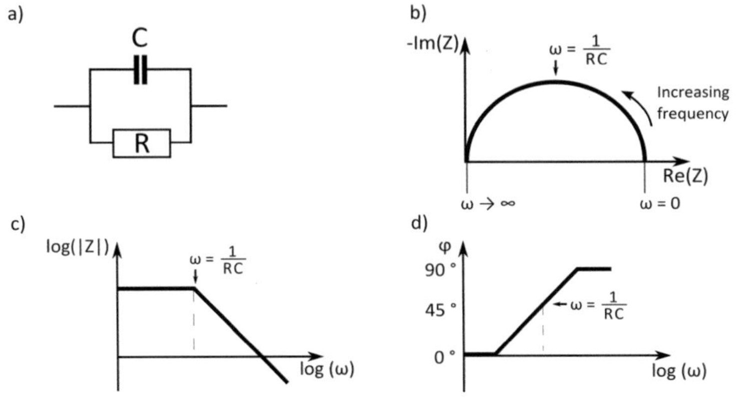

Figure 3 – Comparison of Nyquist and Bode plots for representing the circuit from the resistor in parallel with a capacitor seen in a. b) Nyquist diagram consisting of the imaginary and real part of the impedance over frequency. c) Bode magnitude plot. d) Bode phase plot.

An exemplary Nyquist plot is found in Figure 5. The curve shown is a result from a frequency sweep starting from the upper right side and increasing over the curve to lower left end. This peculiar curve shape will reoccur throughout this work and can be easily explained using an equivalent circuit (see Figure 6) which can be used to describe a system analytically. This modeling technique based on electric circuit components allows for very accurate models from which analytical parameters can be extracted. Table 1 lists the most commonly required elements to describe a biochemical system, namely ohmic resistance, capacitance, constant phase element (CPE) and Warburg impedance. These elements are classified according to their frequency dependence and respective characteristic phase angle. In this table R stands for the ohmic impedance, j is the imaginary unit, ω the angular frequency, C_C is the capacitance, Q is a model parameter analogous to a capacitance, α the transfer coefficient which can change the behavior of the CPE from perfect ohmic resistance ($\alpha = 0$) to perfect capacitor ($\alpha = 1$), σ is the

Warburg coefficient, $\bar{R}$ is the molar gas constant, T is the temperature, n is the number of electron involved, A is the electrode surface area, F is the Faraday constant, D is the diffusion coefficient and K is the concentration of the electroactive species. The subscript "$_O$" and "$_R$" differentiate between the oxidative and reductive species.

Impedance element	Definition	Phase angle	Frequency dependence
R	$Z = R$	0°	No
C	$Z = \frac{1}{j\omega C_C}$	-90°	Yes
CPE	$Z = \frac{1}{Q(j\omega)^\alpha}$	-α·90°	Yes
Warburg [25]	$Z = \frac{\sigma}{\sqrt{\omega}}(1-j)$ $\sigma = \frac{\bar{R}T}{n^2F^2A\sqrt{2}}\left(\frac{1}{\sqrt{D_O}K_O} + \frac{1}{\sqrt{D_R}K_R}\right)$	-45°	Yes

Table 1 – Most commonly used biochemical impedance elements used in the modeling of an electrochemical system.

The combination of these primary elements in parallel or series gives rise to equivalents circuits used to approximate experimental impedance data. For systems which have an electrolyte in contact with an electrode the Randles' cell model is the most commonly used analytical model. This system comprises the solution resistance R_S, charge transfer resistance R_{CT}, double-layer capacitance C_{DL} and Warburg impedance W as seen in Figure 6. The ionic solution resistance accounts for ohmic losses and can be defined as:

$$R_S = \frac{l}{A}\rho \tag{2}$$

Where l is the distance between the electrodes, A is the electrode surface area in contact with the solution and ρ is the specific resistance which varies with the redox concentration. The double layer capacitance results from the fact that the ions on the electrolyte solution are attracted to the charged electrode surface (see Figure 4). The charged ions and the electrode surface are separated by a very small insulating layer consisting of molecules from the solvent (e.g., water). The outer Helmholtz plane is considered the thickness of the double-layer capacitance (see *a* in Figure 4).

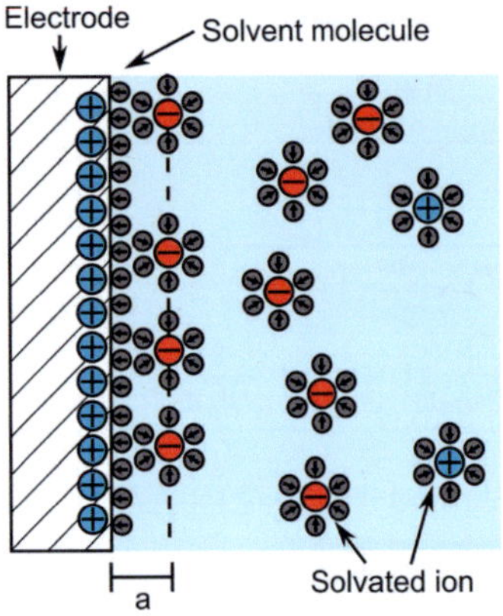

Figure 4 – Schematic of an electrode in contact with an electrolyte solution. Applying a positive potential to the electrode leads to an ion separation, the anions are attracted to the electrode surface. These positive and negative charges are separated by insulating solvent molecules, giving rise to the double-layer capacitance. a) Outer Helmholtz plane.

The double-layer capacitance can be calculated using the following equation, where ε_{DL} is the dielectric constant of the interface layer, *A* the electrode surface area and δ_{DL} the double-layer thickness:

$$C_{DL} = \frac{\varepsilon_{DL} A}{\delta_{DL}} \quad (3)$$

The constant phase element (CPE) models the imperfect behavior of the double-layer capacitance (electrode roughness, porosity, etc.) by adding the transfer coefficient α. The charge transfer resistance represents the intrinsic limitation of the current between electrode and solution by the kinetic rate constant and can be expressed by the following equation in equilibrium and considering an equal concentration of the reductive and oxidative species:

$$R_{CT} = \frac{\bar{R}T}{F^2 A\lambda K} \tag{4}$$

Where $\bar{R}$ is the molar gas constant, T is the temperature, F is the Faraday constant, λ is the kinetic rate constant of the electrochemical reaction, and K is the concentration of the electroactive species. The complete deduction of this formula can be found in [27].

The Warburg impedance describes how the diffusion of the redox probe toward the electrode can influence the current flow and consequently the impedance. Its characteristic frequency-independent 45° line can be explained by its equation found in Table 1, the term $(1 - j)$ implies that the real and imaginary components have the same amplitude, which consequently justifies the 45° angled line (see Figure 5). It is important to remember that a Nyquist diagram shows the negative reactance value, which puts the 45° angled line on the first quadrant.

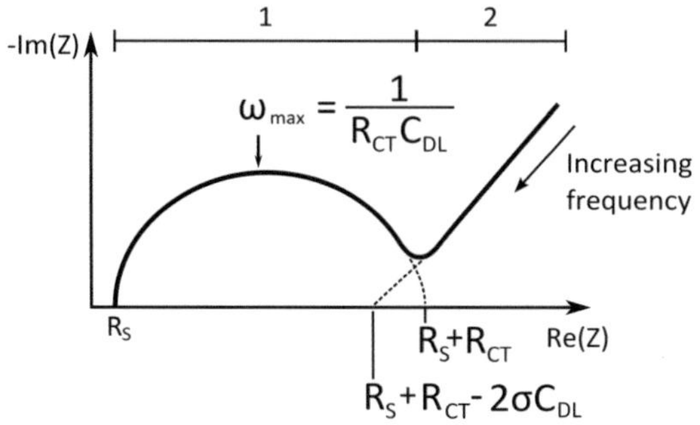

Figure 5 – Nyquist plot of the Randles' cell. This curve is obtained from a frequency sweep with increasing frequency from right to left along the curve. Kinetically controlled processes are predominant in region 1 whilst diffusion phenomena dominate the impedance behavior in region 2.

One can easily confirm that this equivalent circuit (see Figure 6) matches the curve seen in the previous figure. R_S will represent a constant positive offset on the real axis of the Nyquist plot because it represents the solution resistant. At low frequencies $(\omega \rightarrow 0)$ the double-layer capacitor impedance $(Z = 1/j\omega C_{DL})$ tends to infinity and therefore the upper parallel branch will be removed from the circuit. This implies that R_S and R_{CT} are connected in series, thus further shifting the curve to the right. This point represents the rightmost part of the semicircle seen in Figure 5. The Warburg impedance element now dominates the curve behavior as a straight line in a 45° angle, representing the diffusion processes (see region 2 in Figure 5). At high frequencies $(\omega \rightarrow \infty)$ the capacitor impedance tends to zero, thus short-circuiting the lower parallel branch. This means that only the solution resistance will have an effect and represents the leftmost part of the curve.

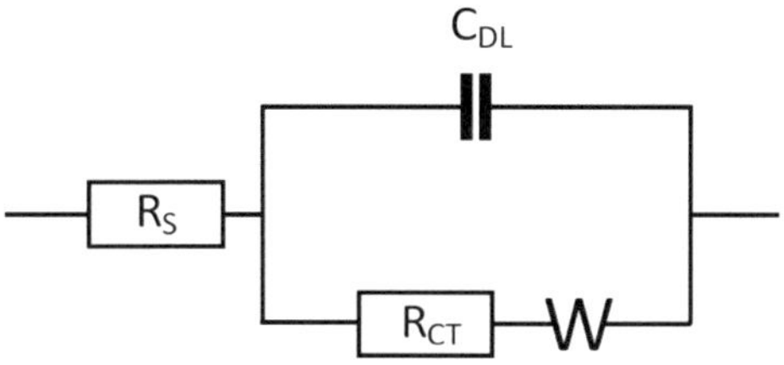

Figure 6 – Randles' equivalent circuit [28] of an electrode in contact with electrolyte. This circuit is the most commonly used model to represent the curve in Figure 5.

The semicircle behavior results from a simple resistor and capacitor in parallel and has its highest point at:

$$\omega_{\max} = \frac{1}{R_{CT} C_{DL}} \quad (5)$$

The parameter extraction from the Nyquist diagram is simple, R_S can be directly obtained, $R_S + R_{CT}$ can be found by extrapolating the semicircle. C_{DL} can be determined by finding the highest point of the semicircle and applying (5). To calculate *W*, the value σ is required. This is done by extrapolating the 45° Warburg-limited line until it meets the horizontal axis and assumes the value:

$$R_S + R_{CT} - 2\sigma C_{DL} \quad (6)$$

2.1.4 Electrochemical cell

For EIS, several electrode configurations are commonly used, the most simple one requires only two electrodes (see Figure 7 a). The working electrode (WE) represents the transduction element during an assay, therefore all biochemical surface modifications will have to be applied to it. The reference electrode (RE) should not be modified during the reaction and has to remain at a fixed potential (E_r). When no voltage is applied between those electrodes each of them assumes a specific equilibrium potential. However, if an external voltage (V) is applied the potential of the WE shifts whilst the potential of the RE remains unaffected. Additionally, an electrolytic current i flows between the electrodes and since the electrolyte solution has an intrinsic resistance (R_S) a voltage drop across the solution (ohmic drop) will result. This amounts to $i \cdot R_s$. Consequently, the potential of the WE (E_w) can be defined as:

$$\begin{aligned} &V = E_w - E_r + i \cdot R_s \\ &\Rightarrow E_w = -V - E_r + i \cdot R_s \end{aligned} \tag{7}$$

This equation shows that the WE potential can be tuned by changing the applied voltage if i is in the range of microamperes and R_S reasonably small. If a constant potential or a low amplitude alternating voltage is applied, the measured current will be approximately constant and therefore the ohmic drop can be easily corrected. The ohmic drop value becomes a critical problem when the voltage-current curve, also known as cyclic voltammogram, of the cell has to be obtained. The high amplitude of the varying voltage applied will lead to unpredictable variations of the WE potential. Introducing a counter electrode (CE, see Figure 7 b) and replacing the voltage source by a potentiostat will enable the flow of the electrolytic current exclusively between the WE and the CE. This separation will guarantee that the equilibrium at the RE is not disturbed and E_r is adjusted with no interference from the ohmic drop.

A potentiostat is a device which controls the potential of the working electrode with respect to the reference electrode by adjusting the current between the working and counter electrode. The potential between WE and RE amounts to the value specified by the user.

The electrode materials vary according to their function. Counter and working electrodes have to be good electrical conductors and chemically stable. Therefore, gold, carbon and silicon compounds are commonly used, depending on the analyte [29]. Reference electrodes are required to have a stable and well-known potential, this is normally achieved by having a redox system with constant concentration of the reducing and oxidizing species, e.g., Ag/AgCl reference electrode whose overall reaction can be describe as:

$$\begin{gathered} Ag^{+} + e^{-} \leftrightharpoons \mathrm{Ag}(s) \\ \mathrm{AgCl}(s) \leftrightharpoons Ag^{+} + Cl^{-} \end{gathered} \tag{8}$$

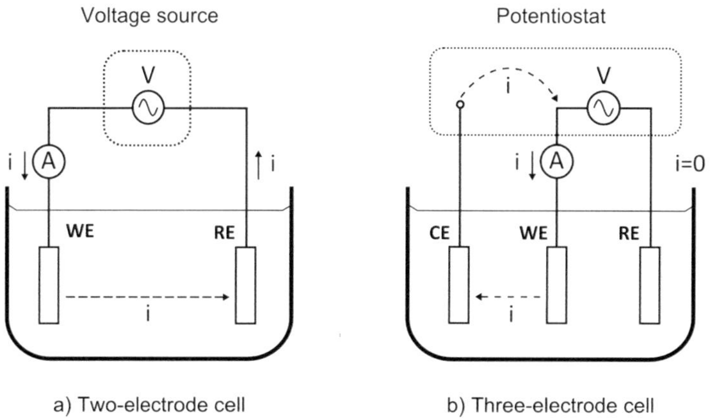

Figure 7 – Schematic diagram of two possible configurations for an electrochemical cell. a) Two-electrode cell configuration with working electrode (WE) and reference electrode (RE) and a voltage source. b) Three-electrode configuration with the addition of the counter electrode (CE), the voltage source is replaced by a potentiostat.

2.1.5 Cyclic voltammetry

Cyclic voltammetry is an electroanalytical technique which studies electroactive species by cycling the potential (see Figure 8 a) of the working electrode against the reference electrode while measuring the resulting current between the working and counter electrode which are shown in a current over potential plot (see Figure 8 b). A triangular potential cycle with amplitude $E_{max} - E_{min}$ and period T is seen in Figure 8 a. The marked points are correlated to the cyclic voltammogram in Figure 8 b. Starting at I with a potential chosen such that the initial electrolysis of the redox species is avoided, the potential is decreased until II. At this point the electrode is sufficiently negative to give rise to the cathodic current which peaks at III with i_{PC} representing the cathodic peak current at the potential of E_{PC}. The potential continues to decrease until IV. The cathodic current decreases because of the depletion of the reducing species. As the potential now increases to V the current reverses and the oxidation process starts. Its peak is reached at VI with the anodic peak current i_{PA} at the potential of E_{PA}. The anodic current starts to decrease due to the depletion of the oxidizing species until the potential reaches the initial value at VII.

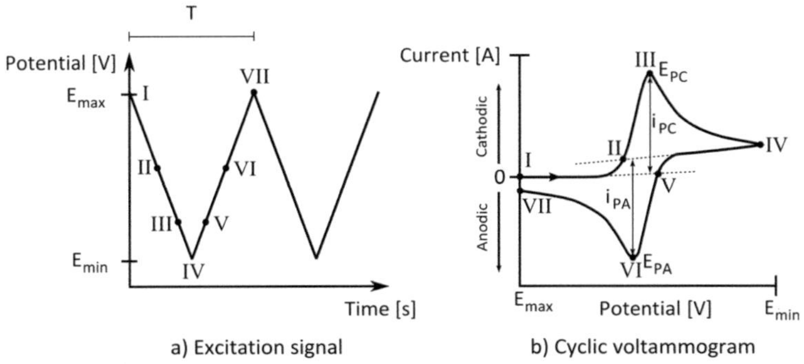

Figure 8 – Excitation and typical cyclic voltammetry curves. a) Triangular excitation signal with amplitude E_{max} - E_{min} and period T. b) Exemplary cyclic voltammogram where i_{PC} and i_{PA} represent the cathodic and anodic peak current and E_{PC} and E_{PA} stand for the cathodic and anodic peak voltages.

2.1.6 Immunoassay

Immunoassay can be defined as a biochemical test which determines the presence or concentration of a specific molecule (also called analyte) in a sample through the use of antibodies or similar binding ligands with high biochemical affinity and specificity. A biosensor used during an immunoassay is called immunosensor. These sensors can detect and quantify a wide range of analytes, e.g., relevant medical diagnostic markers such as proteins, DNA and hormones, illegal and therapeutic drugs as well as bacteria and environmental pollutants such as pesticides.

Immunoassays can be classified into two categories, indirect and direct. Indirect assays can only detect an analyte once an additional marker, normally a radioactive, fluorescence or visible light label, has bound to the immobilized analyte. Typically a primary ligand first binds to the analyte or to a substrate. The detectable label, e. g., in form of a secondary ligand is subsequently bound to the immobilized analyte or to the primary ligand. The most commonly used indirect assay reported in literature is ELISA [17, 18]. One of its many assay formats (indirect, sandwich and competitive) is the so-called indirect assay. It consists of non-covalently immobilizing the analyte to a solid surface followed by the addition of a solution containing the respective ligand (e.g., antibody) which is, in most cases, labelled with an enzyme (e.g., alkaline phosphatase, glucose oxidase, etc.) that reacts with another solution (e.g., para-nitrophenylphosphate) and generates a color change. A stronger color change indicates a higher concentration of the analyte. This color change can be quantified by colorimetric methods. Indirect assays like ELISA are very sensitive but have the disadvantage of being time-consuming, costly and requiring labelled ligands, thus rendering this assay type unsuitable for real-time measurements. Furthermore, the analytes used in sandwich assays, one of the most used assay format, must have two binding sites which is not the case for numerous analytes.

Direct immunosensors are able to detect the biochemical changes during the immune complex formation directly. Common examples of these types of biosensors include gravimetric methods, namely quartz crystal microbalance (QCM) and surface acoustic wave (SAW) [30]. The most commonly studied label-free electrochemical transduction principles are potentiometric, amperometric and impedi-metric [31]. Direct optical methods are mainly based on refractometry (e.g., surface plasmon resonance - SPR) or reflectometry [32]. Finally, calorimetric methods belong to the most studied thermal transduction techniques [33]. Most of these sensors detect the change of surface properties resulting from the binding event.

2.1.7 Impedimetric biosensors

Amongst the multitude of biosensor types available nowadays biosensors based on impedance spectroscopy are particularly well-suited for the direct detection of binding events on the transducer surface. This method requires no markers and can be used in a real-time application. Impedimetric biosensors can be divided into two categories, non-faradaic and faradaic. The latter requires the flow of current through the electrode-electrolyte interface whilst the former only relies on the charging and discharging of the double-layer capacitance.

A non-faradaic biosensor can be understood based on Figure 6. The lack of current flow through the electrode-electrolyte interface implies that the charge transfer resistance assumes a very large value, thus removing the lower branch of the figure from the circuit and resulting in the circuit from Figure 9. The main disadvantage of non-faradaic EIS is the lack of kinetic information about the biochemical binding reaction which effectively takes place on the surface of the electrode. This lack of information about the binding kinetics reduces the suitability of this configuration for immunosensing significantly.

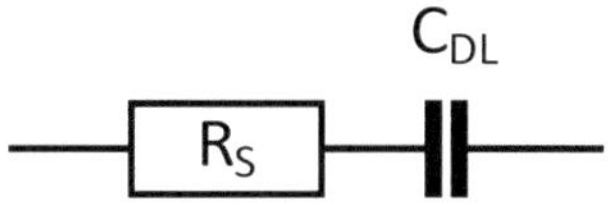

Figure 9 – Randles' equivalent circuit of a non-faradaic electrode in contact with an electrolyte. This circuit is easily obtained from the generic Randles' cell (see Figure 6) when the charge transfer resistance assumes a very large value.

2.2 Antibodies

Antibodies, also known as immunoglobulins (Ig), are Y-shaped proteins (see Figure 10) produced by the white blood cells (B-cells) which can tag or neutralize a specific antigen (e. g., a bacterium or virus). Antibodies have a region called paratope which gives the antibody its specificity towards an antigen (e. g., a transmembrane protein of a bacterium). Antibodies exist in several different configurations (known as isotypes). In placental mammals there are five different isotypes (IgA, IgD, IgE, IgG and IgM) each with different biological properties. This work will focus on IgG as this class is responsible for the majority of antibody-based immunity processes against invading pathogens.

Antibodies consist of heavy and light chains (see Figure 10) which are connected to each other by disulfide bridges. A double disulfide bridge can be found in the so-called hinge region. Different substances can cleave an antibody in different regions. For example papain can separate the antibody into two fragments above the disulfide bridge (Fab) of the hinge region and below it (Fc). Fragments can also be immobilized and used effectively in biosensors [34].

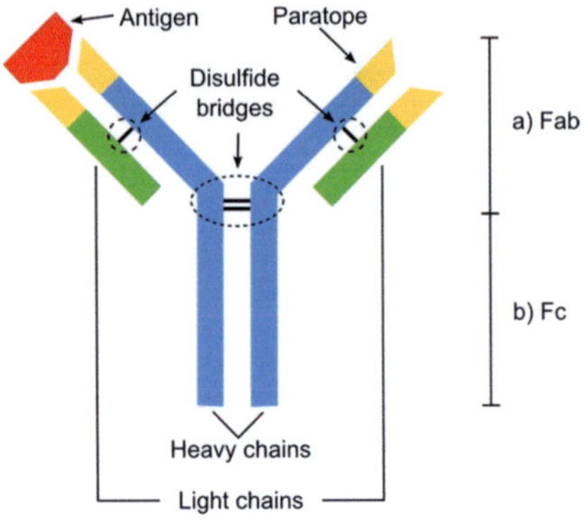

Figure 10 – Illustration of an antibody (also known as immunoglobulin G or IgG). The paratope gives the antibody its specificity against antigens. An antibody consists of a light and a heavy chain which are connected through disulfide bridges.

2.3 Streptavidin-biotin affinity model system

In order to evaluate the performance of the system it is necessary to test it using standard affinity systems used in literature before characterizing the system using cost-intensive antibodies and modified proteins. The most widespread molecule pair used is the streptavidin-biotin model system mainly due to the fact the binding of those two proteins is the strongest non-covalent interactions known in nature with an association constant of about 10^{15} [35]. Furthermore, this protein complex has a high resistance to organic solvents, denaturants, detergents and changes in the pH value.

2.4 Biofouling and biofilm formation

Biological fouling (biofouling) is the accumulation of waterborne bacteria, algae or animals on a technical surface. The bacteria present in such films are embedded in their own extracellular polymeric substances

(EPS) which contain polysaccharides and proteins such as exoenzymes as well as nucleic acids. The EPS grants the bacteria increased resistance to antibiotics and protection against other antimicrobial treatments compared to free-floating colonies of the same cells [36]. Metal, plastic, glass and living tissue are just a few examples of the wide diversity of surfaces affected by biofilm formation. Biofilms are also associated with several infections which range from teeth caries to endocarditis [37]. On a larger scale, biofouling is a problem for industrial systems which are in contact with water. Examples include pipe clogging [38], metal corrosion, efficiency loss of membrane processes [39] and heat exchangers [40] and increased drag of marine vessels [41]. Therefore, several detection methods have been used to better understand the internal mechanisms that govern the biofouling process. Especially relevant are online non-destructive measurement techniques which usually rely on quartz crystal microbalance [42], surface plasmon resonance [43], surface acoustic waves [44], confocal laser scanning microscopy and electrochemical impedance spectroscopy [45].

2.5 Surface modification

This section will discuss several aspects of surface modification. It will discuss why surface motivation is required and highlight the most commonly used methods for doing so with their respective advantages and disadvantages. The most important techniques are described in Figure 14 and will be briefly discussed in the following sections. Furthermore, a brief overview of the surface chemistries required for these methods will be given as well as a discussion of the activation procedures for binding of antibodies to functional groups by means of active-ester chemistry. Finally, an introduction to the mechanism of conductive polymers will be elucidated.

2.5.1 Motivation

The sensitivity of a biosensor depends directly on its ability to detect one analyte specifically whilst blocking the detection of all other substances and molecules present in the sample (see Figure 11). Surface modification plays a very important role as it directly influ-ences the chemical properties of the biosensor surface and therefore its ability to only allow specific (antibody-ligand) inter-actions with the surface and blocks non-specific adsorption.

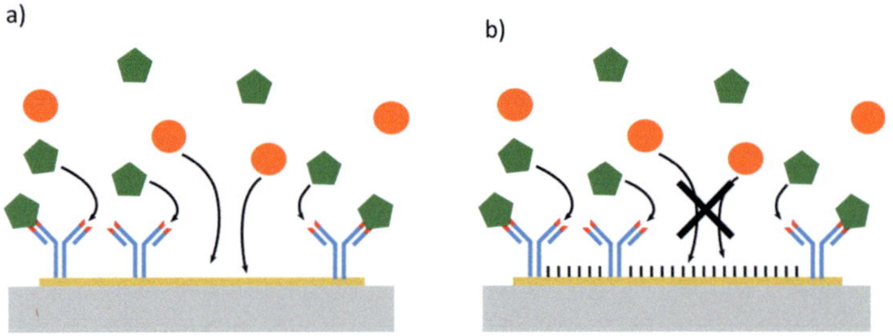

Figure 11 – Schematic diagram of nonspecific adsorption assays on two different surfaces. a) Unprotected surfaces suffer from nonspecific adsorption which leads to measurable (false positive) signals. b) Protected surfaces hinder nonspecific adsorption.

2.5.2 Specific vs nonspecific binding

The ability to detect a specific analyte whilst suppressing the (non-specific) detection of other molecules is of vital importance for immunosensors. Nonspecific binding leads to an increase in false positives rates which dramatically reduces biosensor sensitivity. In the case of biofilm detection there is no need to detect a single analyte amongst thousands of molecules present in the extracellular polymeric substances. As the biofilm colonizes the electrode surface it will hinder the electron flow between the counter and working electrode (see Figure 7 for details on the electrode configuration). This decrease in current or increase in impedance will be an indicator for the biofilm thickness.

2.5.3 Ideal surface modification

As previously explained, an ideal surface modification technique enables the highly efficient binding of the desired ligand to the surface without affecting the ligand affinity to its respective analyte. Furthermore, the surface modification blocks completely the nonspecific adsorption of other molecules. The surface modification required for this work has an additional constraint: It must be of low impedance. The lower the initial impedance is, the higher the achievable sensor sensitivity will be.

Having a surface modification which shows highly efficient ligand binding is not as critical as fabricating a surface modification which blocks completely the nonspecific adsorption of undesired molecules due to the wide availability of chemicals with different functional groups (e.g., carboxyl, amino, etc.) and the existence of robust and established coupling methods (e.g., active ester). The second point represents a challenging task which has been researched intensively for different application fields such as peptide array sensors [46], biosensors [47], antibacterial coatings [48], etc. Most techniques achieve this suppression of nonspecific adsorption by coupling polyethylene glycol (PEG) to the surface of interest. PEG's high hydrophilicity and flexibility have been attributed to its known repellency [49]. The downside of this approach for impedimetric biosensors is the low conductivity of this molecule, especially at high molecular weights (which better hinder the nonspecific adsorption). Therefore small molecules were investigated throughout this work. These molecules had to be highly hydrophilic and should allow convenient coupling to the sensor surface. Tris(hydroxymethyl)-aminomethane (TRIS, see Figure 12 for the chemical structure) fulfils all these requirements. It features three hydrophilic alcohol groups which render the substance very hydrophilic. In addition, the primary amine can be easily coupled to carboxyl groups by active ester chemistry which allows for convenient surface coupling.

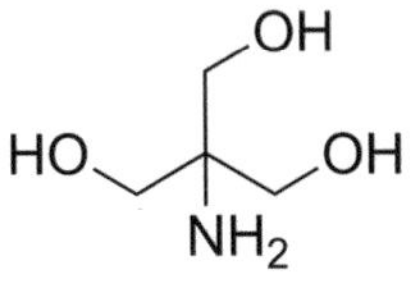

Figure 12 – Chemical structure of tris(hydroxymethyl)aminomethane (TRIS).

As stated, it was deemed very important to generate surfaces with the highest possible conductivities in order to allow the lowest possible surface impedances. Completely functionalizing a surface is not an ideal approach in this case since the immobilized ligands (e.g., antibodies) are not highly conductive (see Figure 13) thus resulting in high-impedance layers. These layers do not allow sensitive impedance spectroscopy measurements. Furthermore, completely covering a surface with ligands could sterically hinder the binding of the respective analyte.

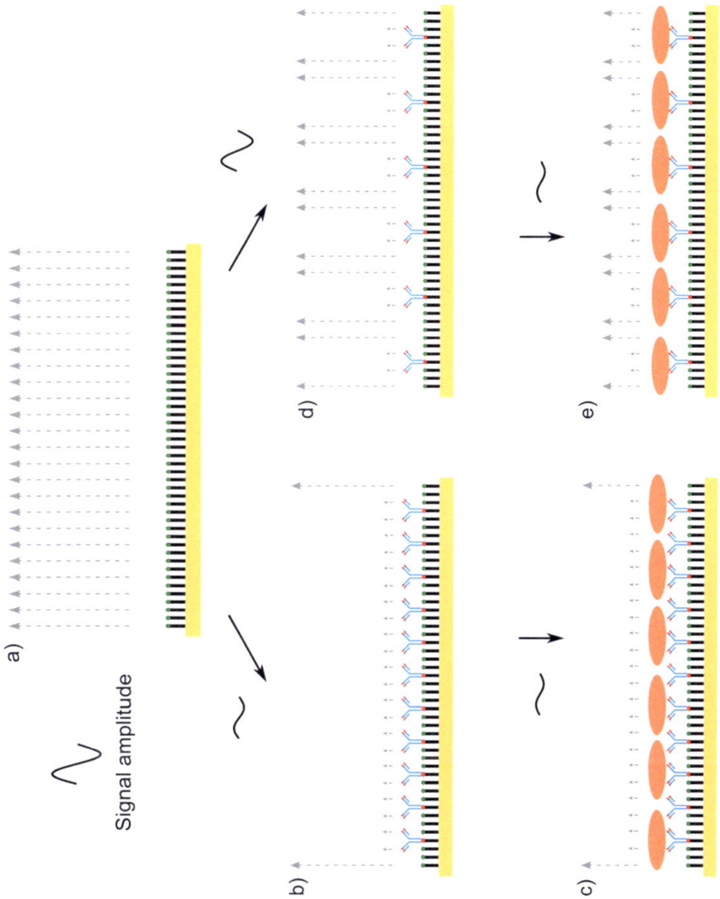

Figure 13 – Schematic illustration of two different ligand coupling strategies on the pre-modified surface of an electrode (a). The gray arrows represent the local current amplitude and sinusoidal curve represents the overall signal amplitude. b) The comprehensive coupling of ligands to the pre-modified surface. c) The binding of the analyte to the ligand results in only a slight current reduction. d) The sparse coupling of ligands to the pre-modified surface. e) The binding of analyte significantly reduces the overall current.

Figure 13 illustrates two different ligand coupling strategies on the pre-modified (e.g., SAM) surface of an electrode (see Figure 13 a). The gray arrows represent the local current amplitude and the sinusoidal curve depicts the overall signal (current) amplitude. If the pre-modified electrode surface is comprehensively covered with bound ligands this will lead to a drastic current reduction as the ligands are bulky and insulating (see Figure 13 b). The subsequent binding of the analyte to the ligand is responsible for a slight current reduction (see Figure 13 c). This leads to a small surface impedance difference prior and after this binding event. The other coupling strategy is based on sparsely coupling ligands to the pre-modified surface which in turn leads to a moderate overall current decrease (see Figure 13 d). The subsequent binding of the analyte leads to a significant reduction of the overall current (see Figure 13 d). This scenario allows for much higher sensor sensitivity as the signal difference prior and after the analyte binding event is much higher than in the previous scenario. This signal difference will be another very important factor determining the sensor sensitivity.

One has to keep in mind that the schematic diagrams shown in this work do not represent the substances in their original scale for didactic purposes. Taking as example allyl mercaptan, a short chain unsaturated thiol compound which will be used as a surface modification substance for the method based on photobleaching. The molecular weight of allyl mercaptan is approximately 1000 times smaller than the molecular weight of bovine serum albumin (BSA) and IgG, 4 times smaller than the molecular weight of biotin and 2 smaller than the molecular weight of TRIS. The comprehensive immobilization of ligands to the pre-modified surface will not necessarily reduce the nonspecific adsorption as most likely will reduce the sensor's sensitivity. Far more decisive is the quality underlying pre-modified surface.

2.5.4 Immobilization techniques

Antibodies are very often used as recognition elements for specific target antigens in immunosensors. Antibodies (see Section 2.2) possess two binding paratopes. In order for this protein to retain its function it is required that the binding sites remain unmodified and unobstructed. These may seem trivial requirements but ensuring them has required enormous effort from researchers. Several different approaches to correctly immobilize antibodies have been investigated over the years, ranging from self-assembled monolayers [50], conductive polymers [51], entrapment in a polymer matrix [52], chemical grafting of azo compounds [53] to affinity binding using suitable functional groups [54]. Each method possesses distinct advantages, disadvantages and limitations with respect to the substrate material and desired physicochemical properties (i. e., surface conductivity, transparency, chemical stability, etc.). Furthermore, some of the methods enable the localized immobilization of antibodies only on certain materials such as, e. g., gold electrodes or glass substrates (see Figure 14).

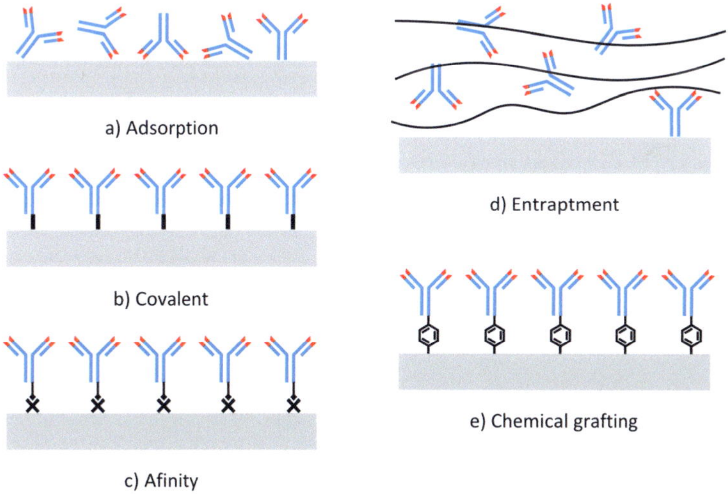

Figure 14 – Schematic representation of antibody immobilization techniques.

Adsorption

Physical adsorption onto a solid surface is one of the simplest methods to immobilize antibodies. The physical interaction of the antibodies with the surface gives rise to nonspecific weak bonds, i.e., hydrogen bonds, electrostatic interaction and Van der Waals' forces with the surface. A major advantage of this method is that no additional reagents are required, resulting in very simple immobilization procedures. Consequently, adsorption is easily carried out, cheap and normally does not hinder the binding affinity between antibody and antigen. On the other hand, the weak bonds tend to be easily disrupted by changes in pH, temperature and presence of surfactants thus changing the surface coverage over time. ELISA-based biosensors which use an indirect assay format usually rely on this technique for analyte immobilization [55]. In general, purely adsorbed antibody layers are not stable in operation and thus should be avoided.

Covalent immobilization

Covalent immobilization requires the presence of functional groups on the surface to which the analyte specific molecules will be immobilized. The main advantage of this method is the assured orientation and thus the guaranteed availability and accessibility of the binding sites of the antibody. On the other hand, this method requires the use of additional reagents. The choice of coupling chemistry must ensure that the binding sites remain unhindered and intact thus retaining the antibody's affinity. In addition to choosing the correct functional group on the antibody, the surface to which the antibody is to be immobilized must be prepared accordingly. There are several methods for modifying a surface with functional groups such as, usage of self-assembled monolayers (see Section 2.5.5 for details). Carboxyl functional groups can easily be activated to react with amino groups forming a peptide bond by using active-ester chemistry (see Section 2.5.6).

Affinity interaction

This method is based on the affinity of two additional molecules, one linked to the substrate and the other to the molecule which should be immobilized. The most commonly used pair of molecules for this method is biotin (a vitamin) and streptavidin (a protein with four identical binding sites with a very high affinity to biotin) [54] as well as avidin. This method has three main advantages. Firstly, the large availability of pre-functionalized antibodies and other molecules with biotin or (strept-)-avidin tags which lend themselves directly to immobilization. Secondly, no additional reagents are required for the immobilization. Thirdly, multiple binding sites may be used on one ligand thus amplifying the signal readout. On the other hand, the affinity bond of biotin and streptavidin can be denatured by surfactants, pH and temperature. More critically for EIS and gravimetric methods, the additional layer of affinity molecules reduces the layer conductivity drastically, thus decreasing the sensor's sensitivity.

Entrapment in a polymer matrix

The process of mixing the desired ligand (usually an antibody) with a monomer or diluted polymer solution and subsequent polymerization or solidification of the polymer is referred to as entrapment [52]. By performing this process the ligand will be embedded, or entrapped, in the polymer matrix. This is a very simple method to immobilize analyte-specific molecules and requires no additional chemical modifications. On the other hand this method cannot assure that the binding sites of the entrapped molecules are oriented or available for binding. Furthermore, the polymerization process may denature the protein.

Chemical grafting

Chemical grafting is a procedure through which a surface can be modified by addition of small molecules or functional groups. One of the most widely used methods is grafting of aryl diazonium salts to several different surfaces [53] based on azo chemistry (see Section 2.5.5 for details). Different aryl diazonium salts containing different functional groups enable a wide range of different functionalizations. The disadvantage of this method is the complexity of the azo chemistry which requires several reagents and permanent temperature control throughout the process.

2.5.5 Covalent surface chemistry

Self-assembled monolayer

Self-assembled monolayer (SAM) is one of the most simple surface modification methods [50]. SAM-forming molecules consist of three chemical blocks or functional groups (see Figure 15): a thiol group which interacts strongly with gold, a carbon backbone which is responsible for the self-organization and coverage density and a functional group to which analyte-specific molecules (e.g., antibody) can be bound. Applying this layer to a gold surface is accomplished by dissolving the SAM-forming molecule in an appropriate solvent, e.g., ethanol and incubating this solution on the gold surface overnight. The SAM-forming molecule will self-align on the surface and create a homogeneous monolayer. The length and type of the carbon chain will greatly determine the electrical properties of the SAM in question.

Using SAM with impedimetric sensors poses an additional challenge. Besides being as homogenous and dense as possible, a SAM suitable for EIS must also be highly conductive. Low impedance of the SAM layers allows for high overall sensor sensitivity. Electrically insulating SAMs will significantly decrease the sensitivity.

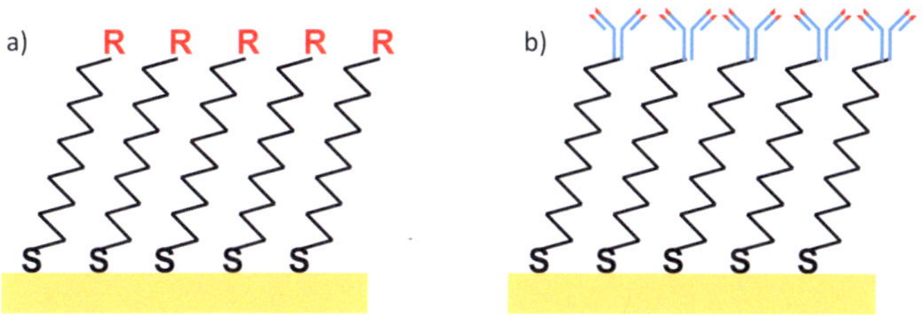

Figure 15 – Schematic representation of self-assembled monolayers (SAM) on a gold substrate. a) Initial state after SAM formation. b) The immobilization of ligands can be carried out easily using different activation techniques, e.g., active ester chemistry. The substances shown in the diagram are not in original scale for didactic purposes.

Chemical grafting

Aryl diazonium salts (see Figure 16 a) are widely used compounds in organic chemistry due to their ability of introducing halogens, CN, OH, H, etc. into aromatic rings [53]. Additionally, the reaction of diazonium salts with aromatic molecules generates a very important class of dyes known as azo dyes. The grafting process is normally carried out in acidic media using sodium nitrite. The acidic media will deprotonate the primary amine whilst sodium nitrite ($NaNO_2$) will induce the formation of the aryl diazonium salt. This highly instable compound requires a stringent temperature control as it deteriorates at temperatures above 5 °C. Sodium nitrate will reduce the aromatic amine (see Figure 16 b) to the highly unstable diazonium cation (see Figure 16 c). Applying a potential to this complex promptly splits off nitrogen thus generating aryl radicals (see Figure 16 d). Rapid nitrogen gas release can be dangerous which is why the whole process has to be tightly temperature controlled and conducted below 5 °C. These radicals can bind to the surface of the cathode electrode (see Figure 16 e).

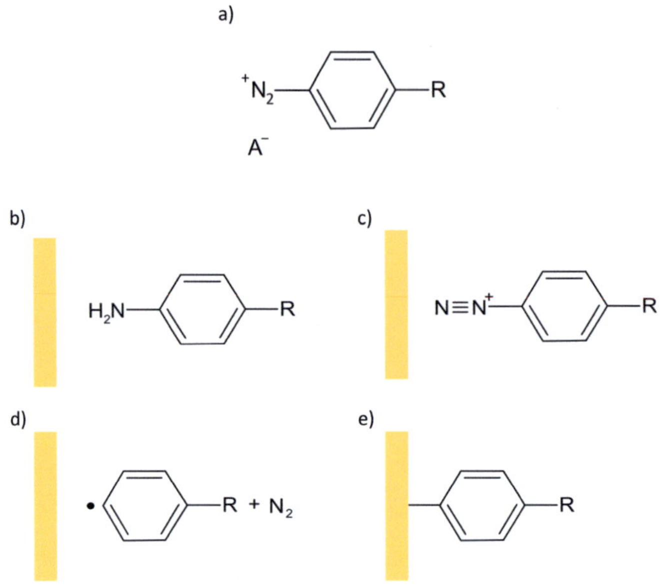

Figure 16 – Grafting reaction process for aryl diazonium salts. R represents a functional group and A^- the counter anion.

Photobleaching

Photobleaching is a photochemical process which fluorophores undergo when they are excessively irradiated causing their permanent loss of fluorescence. It is an undesired phenomenon for microscopy methods as it leads to strong emission intensity losses. Holden and Cremer have shown in 2003 [56] how to use this phenomenon to immobilize ligands coupled with a fluorophore to BSA-coated surfaces. The overexposure of the fluorophores to their absorption wavelength leads to its destruction and the generation of short-lived radicals which can be used to couple ligands to the BSA surface. Curtis *et al.* have shown the successful coupling to unsaturated compounds such as methacrylate silanes [57]. This method enables the creation of protein patterns using a process similar to lithography which works in visible light and with water-based solutions [58]. Using visible light is advantageous for biological applications as UV light damages proteins and DNA.

In this work a novel photobleaching method has been used in combination with SAMs. Allyl mercaptan is used as SAM-forming reagent which creates an unsaturated surface providing groups for the coupling of photobleached dyes (e.g., fluorescein).

During this work fluorescein (see Figure 17 a) was used because it is inexpensive and widely available as a separate dye as well as conjugated to innumerous proteins and antibodies. Furthermore, when fluorescein is coupled to an isothiocyanate (fluorescein isothiocyanate, FITC) group (see Figure 17 b) it can promptly and effectively react with primary amines. This allows for the easy and fast labeling of proteins and antibodies, e.g., fluorescein-TRIS, fluorescein-BSA, fluorescein-IgG.

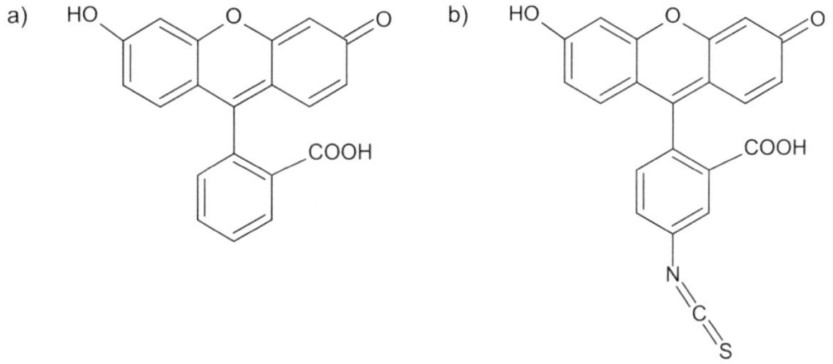

Figure 17 – Chemical formula of fluorescein (a) and fluorescein isothiocyanate (b).

2.5.6 Active-ester chemistry (EDC/NHS)

Active-ester chemistry is a very common technique used for covalent immobilization of analyte specific molecules. These substances promote the binding of carboxyl groups with primary amines via peptide bond formation which are, for instance, responsible for the linkage of protein chains.

Two reagents are required during this process (see Figure 18), a carbodiimide (in aqueous solutions usually 1-ethyl-3-(3-dimethylaminopropyl)-carbodiimide hydrochloride, EDC) and the alcohol forming the ester. Typically, N-hydroxysuccinimide (NHS) is used in biochemistry. Carbodiimides react with carboxylic acids forming an unstable intermediate (O-acylisourea) which then binds to the alcohol (in this case NHS) forming an activated ester prone to nucleophilic attack. NHS is conveniently displaced by a primary amine (a strong nucleophile) which then forms a peptide bond with the carboxylic group. The usage of NHS combined with EDC leads to a much more stable and efficient reaction when compared to one with only EDC [59].

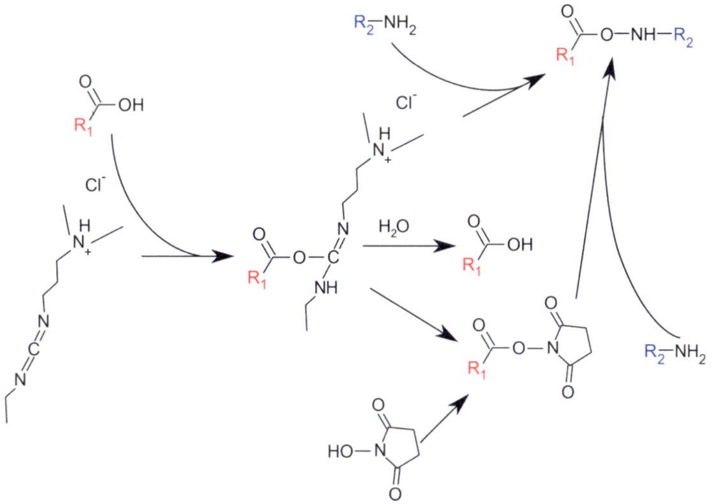

Figure 18 – Reaction schematic for active-ester chemistry using 1-ethyl-3-(3-dimethylaminopropyl)carbodiimide hydrochloride (EDC) and N-hydroxy-succinimide (NHS) to form a peptide bond between a target carboxylic acid and primary amine.

2.5.7 Conductive polymers

Polymers, e.g., plastics, are ubiquitous in every-day life, ranging from low-tech packaging materials, tubing or car parts to high-tech medical implants and high-precision optical elements. These materials are normally considered as electrical insulators. For instance, polyethylene has a conductivity of 10^{-16} S/m which is 23 orders of magnitude lower than copper. The fact that polymers could be used as conductors was only introduced in 1977 by Alan MacDiarmid, Alan Heeger and Hideki Shirakawa [60]. By adding dopants (Br_2, I_2, AsF_5) to polyacetylene they managed to increase the conductivity from 10^{-6} to 10^{7} S/m. This discovery paved the way for a new class of materials: conductive polymers.

In this work, conductive polymers were used as matrices for the immobi-immobilization of antibodies. Polypyrrole (PPy) was chosen as suitable polymer. PPy is one of the most widely studied conductive polymers due to its ease of synthesis, stability and high electrical conductivity [61]. PPy can be synthesized by chemical and electrochemical oxidation in solution. The polymerization process via monomer linkage normally takes place on the 2/5 position (see Figure 19). Pending functional groups are normally attached to 3/4 positions. The presence of functional groups has drawn attention to this material for usage in biosensors since it enables the immobilization of analyte specific molecules, e.g., antibodies and enzymes.

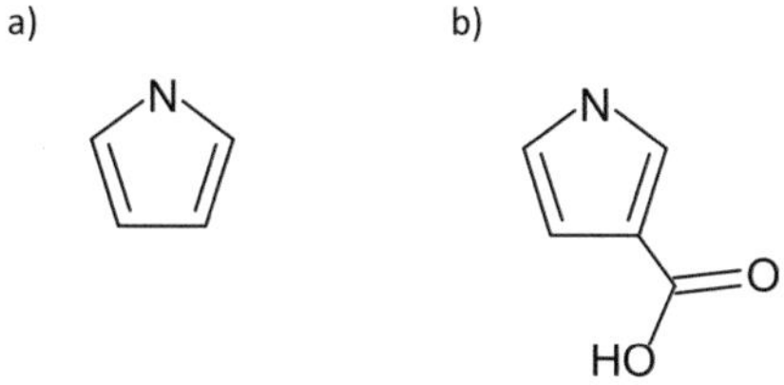

Figure 19 – Monomers used for synthesis of electrically conductive polymers in this work. a) Pyrrole, b) Pyrrole-3-carboxylic acid.

2.6 Photolithography

Photolithography is an optical microfabrication technique which can transfer a structure from a pattern, also called photomask, onto a substrate through a polymeric intermediary layer know as photoresist. Photoresists are light, mainly ultraviolet light (UV), sensitive polymers which can be classified into two categories: positive and negative photoresists. This categorization comes from the changes in solubility of the photoresists in the developer solution after light exposure. The solubility of the positive type photoresists increases whilst the solubility of negative type photoresists decreases.

2.6.1 Types of photoresist

As described, photoresists are light sensitive substances which, depending on the underlying mechanism can be chemically changed permanently. Positive photoresists include AZ photoresist, Poly(methyl methacrylate) (PMMA), Kodak micro positive resist (KMPR), etc. and are mostly used in thin-films (several 10 µm). Negative photoresist comprise, among others, SU-8 which is often used as thick film resists (several 100 µm).

2.6.2 Negative photoresists

One of the most commonly used negative photoresist is SU-8, an epoxy-based oligomer which can be applied in films ranging up to 300 µm in thickness. This is only possible due to its high cross-linking enabled by its 8 epoxy subunits which in turns make SU-8 very stiff once cured. It is therefore suitable for high aspect ratio applications in microelectromechanical (MEMS) systems [62]. On the other hand, due to its high degree of cross-linking it is very difficult to strip SU-8 once it has been cross-linked by UV light exposure.

2.6.3 Positive photoresists

One of the most commonly used class of positive photoresist is the AZ series by AZ Electronic Materials GmbH. These resists consist of a base resin of phenol/formaldehyde (see Figure 20 a), a photosensitive solution known as diazonaphthoquinone (DNQ, see Figure 20 b) and a solvent (in the case of the resist used in this work this solvent was propylene glycol monomethyl ether acetate, PGMEA).

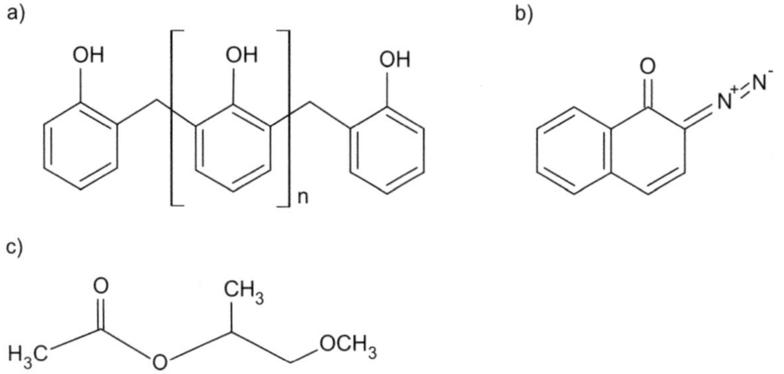

Figure 20 – Chemical components of the photoresist AZ 1512 HS. a) Base resin of phenol/formaldehyde. b) The photosensitive compound diazonaphthoquinone. c) Solvent, propylene glycol monomethyl ether acetate.

The phenol formaldehyde resin is synthesized from phenol and formaldehyde in a molar ratio smaller than 1 (formaldehyde to phenol). Using this ratio these polymers are also known as novolacs. Using under-stoichiometric amounts of formaldehyde prevents the complete cross-polymerization of the resin via methylene bridges. The different ratios of phenol and formaldehyde influence several properties of the photoresist (melting point, developing speed, dark erosion rate, etc.) [63]. The solubility of the base resin in aqueous alkaline media is moderate (around 15 nm/s). After the addition of 25 to 30 m% of DNQ the solubility decreases two orders of magnitude. After being exposed to ultraviolet light (wavelength < 440 nm) DNQ reacts to form a carboxylic acid containing compound. This compound is now 3 to 4 orders of magnitude more soluble in alkaline media, thus being 1-2 orders of magnitude more soluble than the base resin [64]. This means that unexposed regions of the photoresist have a very low solubility in alkaline media whilst the exposed areas have a very high solubility, therefore enabling the creation of a "binary" pattern on the photoresist. PGMEA is the chosen solvent because it has a high boiling point (145°C), low vapor pressure (approximately 5 mbar at room temperature), thus avoiding rapid evaporation while the resist is being applied to a surface, e. g., via spin-coating. Furthermore, PGMEA is one of very few solvents that, even when the photoresist is highly diluted does not promote the agglomeration of the resin base.

Figure 21 describes in more detail the chemical transformation that DNQ undergoes upon ultraviolet irradiation. This process is known as Wolff rearrangement. Initially, the ultraviolet radiation cleaves the Nitrogen branch, releasing molecular nitrogen, and forming an intermediary carbine which readily rearranges into a reactive ketene. The ketene reacts in presence of water resulting in a carboxylic acid [65]. This form shows a much higher solubility in alkaline media than DNQ.

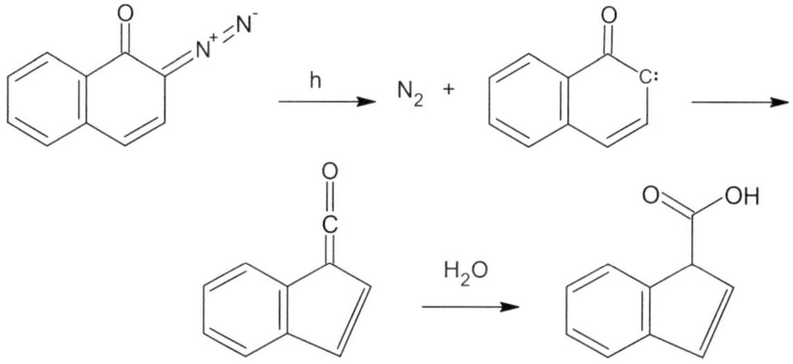

Figure 21 – Wolff rearrangement of DNQ under ultraviolet light.

This photochemical transformation enables the use of this photoresist as a mask directly applied to a surface. Photoresists containing DNQ are also suitable for thick film (> 10µm) processes. This is due to the fact that, once the DNQ has undergone Wolff rearrangement, it becomes transparent to ultraviolet light down to 300 nm, thus enabling the exposure of the underlying layers [66].

2.6.4 Photomask

A photomask is a transparent substrate with opaque patterns which determine the structures transferred to the photoresist layer. Therefore the mask's line resolution will determine the structure resolution on the pattern transferred to the photoresist. There are several methods for fabricating or implementing a photomask. These methods can be subdivided into parallel or sequential methods. The first method is able to transfer the entire pattern of the photomask within one exposure. Typically chromium masks or printed masks are used for this as well as maskless projection lithography. Sequential methods scan the desired surface using, e.g., an electron or laser beam thus serially writing the pattern.

Chromium masks (e. g., fabricated via electron beam direct writing) have a very high resolution but are expensive and time consuming to fabricate. Serial methods require expensive equipment and are time consuming especially if complicated designs are to be written. Therefore, in order to test several electrode designs with low cost methods, printed masks and maskless projection lithography were chosen during the course of this work.

2.7 Gold sputtering

Sputtering is a deposition process in which gas ions (e.g., argon ions) are accelerated in an electric field inside of a vacuum chamber against a target material. This ion bombardment will cause atoms from the target to escape the bulk material which will subsequently be transferred to a substrate, thus forming a film (see Figure 22).

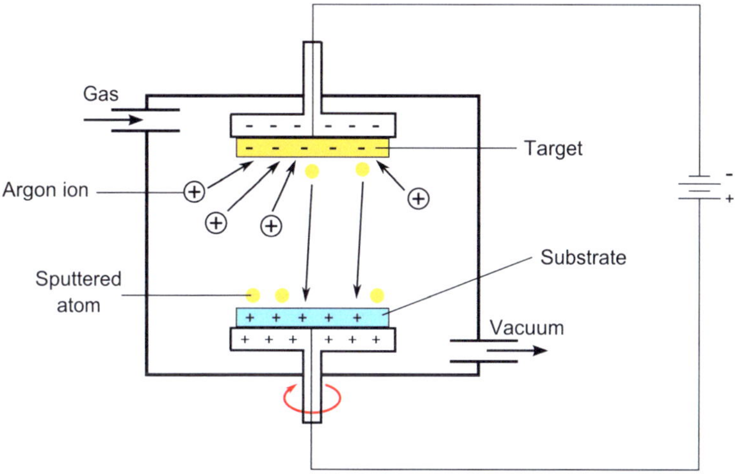

Figure 22 – Schematic diagram of a direct current (DC) sputtering machine. Inert gas atoms are accelerated in an electric field inside a vacuum chamber against the target material. Atoms from target are removed and deposit onto the substrate forming a homogeneous layer.

After the chamber has been filled with the sputtering (neutral) gas under reduced pressure and a DC electrical voltage is applied, free electrons are accelerated away from the target. When they come close to a neutral gas molecule they will drive the outermost electrons off, thus ionizing the gas. The positively charged gas molecules will now be accelerated towards the target material because of the applied electrical field. This target bombardment with charged gas molecules will lead to the removal of atoms from the surface of the target and generate additional free electrons. In the meantime the recombination of ionized gas molecules and free electrons takes place. This phenomenon will bring the free electrons to the ground state, thus resulting in the emission of a photon which gives plasma its color [67]. The atoms removed from the target are deposited onto the substrate as a thin homogeneous layer. Parameters like distance from target to substrate, sputtering current and target material determine the deposition rate.

2.8 Electrode manufacturing process

This section describes shortly the electrode manufacturing process which is schematically described in Figure 23. The first step is to decide which substrate will be used depending on the application, process requirements (chemical resistance to substances used during the fabrication process) and desired mechanical properties (e.g., transparency, porosity, hardness, etc.). The chosen substrate is then cleaned and is ready to be sputtered (see Section 2.7) with the desired metal layer (e.g., gold, copper, titanium, silver, etc.). It is also important to consider the adhesion force between the metal layer and the substrate material. Some substrates require additional adhesion layers, e.g., gold on glass requires a thin titanium or chromium layer. Afterwards, a positive photoresist is deposited on top of the metal layer by spin-coating (see Section 3.1.3), after a short pre-baked interval the photoresist is ready to be exposed. After selecting the correct spectral range of the UV lamp and aligning the mask with respect to the substrate the photoresist is exposed. The resist can then be developed. If a positive resist was used all the exposed areas will be soluble in the developer solution. If a negative resist was used the inverted structure will be created. At the end of the developing step a structured photoresist layer will remain on top of the metal layer. The exposed metal layer can now be etched whilst the photoresist protected areas will remain in place. Finally, the remaining photoresist and, if necessary, the adhesion layer can be removed.

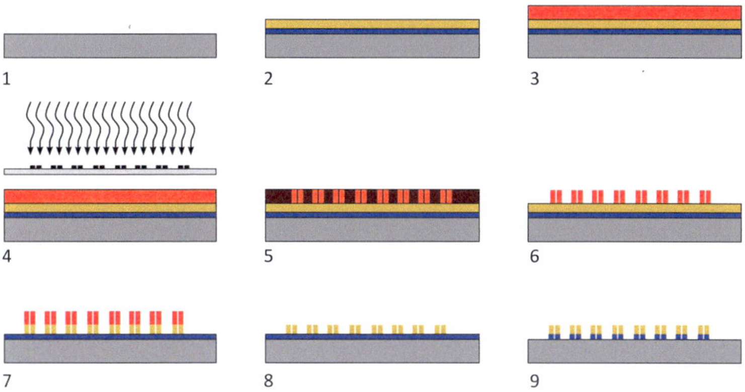

Figure 23 – Schematic diagram of electrode fabrication steps used in this work. 1) Substrates are cleaned. 2) A gold layer and (in case glass is used) a titanium adhesion layer are applied by means of sputtering. 3) Spin coating of a positive photoresist. 4) UV exposure through a photomask. 5) Exposed photoresist areas are removed in the developer solution. 6) After development the metal layer with the structured photoresist remains on the substrate. 7) Gold etching in KI/I_2 solution. 8) Removal of the remaining photoresist. 9) Etching of the titanium adhesion layer (if used) in piranha solution.

2.9 Fluorescence microscopy

Fluorescence and phosphorescence are physical phenomena based on the absorption and subsequent re-emission of light by organic and inorganic substances. Phosphorescence emits light for a long period after the extinction of the excitation light. The emission of light in fluorescence is an almost simultaneous process as it usually takes less than one microsecond delay between photon absorption and emission [68]. Fluorescence was first described by Stokes in 1852 observing that the mineral fluorspar (CaF_2) emitted red light when exposed to UV light. Many fluorescent materials were discovered in the 19th century but it was only during the 1940s that this phenomenon was first used to investigate biological tissue stained with fluorescent chemical compounds (fluorophores). This discovery led to the development of several different detection and microscopy techniques and is today an essential and established biological detection method.

The most commonly used fluorescence microscopy setup is depicted in Figure 24. A light source emits a wide spectrum of light (normally an arc discharge lamp or high-pressure mercury lamp) which is filtered by the emission filter allowing only the light with the excitation wavelength (excitation beam, in blue) of the fluorophore to pass. This light impinges a dichroic mirror which reflects only the excitation beam. This beam is focused through an objective onto the substrate where the specimen is placed. The fluorophore absorbs the incoming light and re-emits longer wavelength light due to internal losses (non-radiative transitions, vibrational losses). The emission beam goes through the dichroic mirror and emission filter. Finally, the light goes through the ocular element and reaches the detector (eye or camera detector).

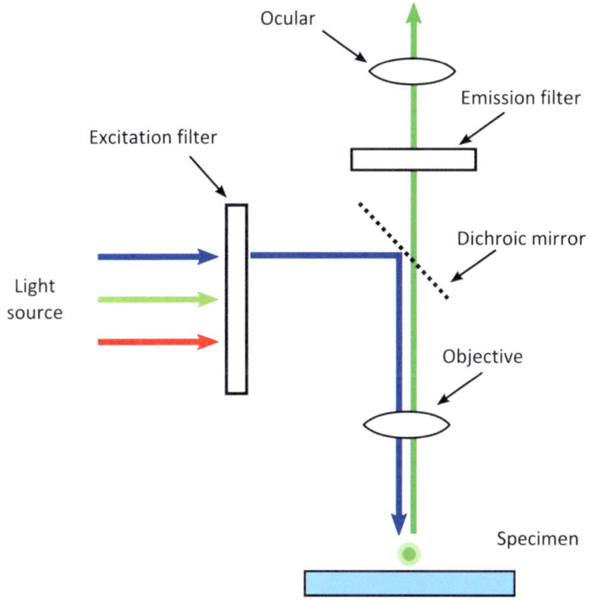

Figure 24 – Schematic diagram of a vertical illumination fluorescence microscope.

2.9.1 Fluorescence microscopy as a complementary technique

As previously mentioned, fluorescence microscopy has become a reference technique for biological analysis. Many new detection methods are compared with results obtained using fluorescence microscopy.

Confirming the results obtained by impedance spectroscopy using fluorescence microscopy can be very valuable and easily accomplished. Proteins and antigens can be modified using fluorescence labels such as fluorescein, cyanine, alexa dyes, etc. This modification does not have a considerable influence on the impedimetric results as long as the molecular size of the fluorophore is much smaller than that of the analyte. Furthermore, fluorescent antibodies can be used to determine the efficiency, homogeneity and density of the immobilized antibody layer. The combination of impedance spectroscopy with fluorescence microscopy can provide important characterization data as well as complementary verification measurements.

2.10 Microfluidics

During the beginning of the 1990's a change in the materials used to make microfluidic chips started to take place. Due to the increasing demand for disposable and cheap microfluidic devices for biomedical and clinical applications polymers were chosen as new materials of choice. Silicon and glass were the first materials used in microfluidic chips but they possess several disadvantages as microfluidic chips material [69]. They are not suitable for large scale production due to the required complex etch steps with dangerous substances (e.g., hydrofluoric acid, HF). Furthermore, they lack gas permeability which renders them unsuitable as cell culturing material; additionally their high stiffness made it impossible to create devices with movable parts such as pumps and valves. Polymers were not only cost-efficient but could also be readily replicated on an industrial scale. The most commonly used process is injection molding which allows for low-cost and high-throughput production.

In the late 1990's the Whiteside's group presented a replication technique for polydimethylsiloxane (PDMS, also known as silicone) from structured silicon molds with feature sizes ranging from 100 µm down to 30 nm [70]. Due to its simplicity and ease of use this process is still predominant in the microchip fabrication process for academic purposes. This technique was also used for the fabrication of the microchannels which house the gold electrodes.

PDMS possesses several undesirable properties for immunosensors, e.g., high oxygen and water vapor permeability, adsorption of hydrophobic molecules and lack of chemical resistance. In order to overcome those disadvantages the PDMS flow cells used in this work were covered by a protective layer of parylene C (see Figure 25) a transparent thermoplastic polymer which is extensively used as coating material for biomedical implants [71]. Parylene C is non-biodegradable and resistant to most chemicals. Thin layers of parylene can be deposited by means of chemical vapor deposition (CVD). Besides its excellent *in vivo* biocompatibility parylene C has been established as a suitable material for culturing cells [72].

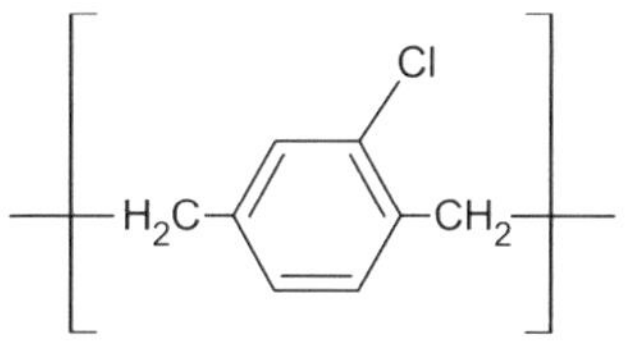

Figure 25 – Chemical structure of a parylene C repeat unit. Parylene C is a transparent and chemically inert thermoplastic polymer.

3 Development and application of the sensor system

This chapter lists the used materials and explains the techniques used to fabricate the electrodes and flow cells. Furthermore, it elucidates the surface modification methods used and how a measurement using EIS and fluorescence microscopy was carried out.

3.1 Electrode manufacturing process

This section describes shortly the electrode manufacturing process which has been shown schematically in Figure 23 and is pictured in Figure 26. The substrate material has to be chosen depending on the application, process requirements (chemical resistance to substances used during the fabrication process) and desired mechanical properties (e.g., transparency, porosity, hardness, etc.). At the beginning of this work plastic substrates (cyclic olefin copolymer, COC) were used but the implementation of more complex surface modifications showed that the gold adhesion on plastic was insufficient. Ceramic (Al_2O_3) was tested afterwards but due to its high porosity it could not be used. Lastly, glass substrates were successfully tested and became the standard substrate material.

The blank substrate (Figure 26 a) had to be sputtered with gold. This step was not required for glass substrates because they were purchased precoated with gold (Figure 26 b). Due to the insufficient adhesion of gold on glass a thin titanium adhesion promoter layer (10 nm) was necessary. Afterwards, a positive photoresist (AZ 1512 HS) was deposited (Figure 26 c) on top of the metal layer by spin-coating, after a short pre-bake the photoresist was exposed (Figure 26 d). After exposure, the transferred pattern can already be seen (Figure 26 e). As a positive resist was used, the exposed areas become soluble in the devel-developer solution (Figure 26 f). The exposed gold layer was then etched using a KI/I_2 solution whilst the areas protected by the photoresist remained unaffected (Figure 26 g). Finally, the remaining photoresist (Figure 26 h) and, if necessary, the adhesion layer (Figure 26 i) were removed in acetone and piranha solution, respectively.

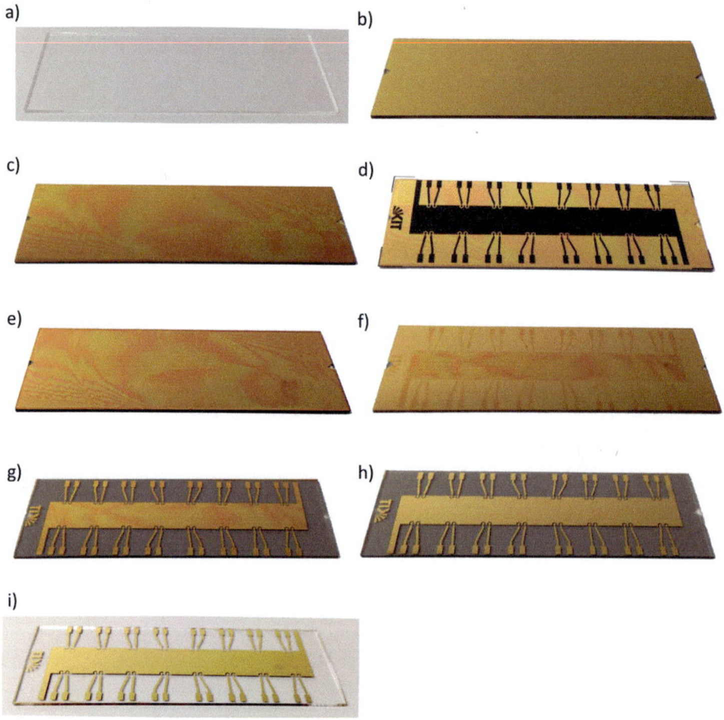

Figure 26 – Pictures of the electrode manufacturing procedure. a) Clean glass substrate. b) Titanium and gold were sputtered. c) Spin coating of a positive photoresist. d) UV exposure through a photomask. e) Exposed photoresist areas can then be removed in the developer solution. f) Substrate after the development of the exposed photoresist. g) Substrate after gold etching in KI/I_2 solution. h) Substrate after removal of remaining photoresist in acetone. i) Substrate after etching of the titanium adhesion layer in piranha solution.

3.1.1 Mask fabrication

As described in Section 2.6.4 there are several ways to implement a photomask. In this work printed masks and maskless projection lithography were used. Initially, masks were printed using a laser printer (see Figure 27 a) but due to its low feature resolution limit of 100 µm and insufficient edge resolution an alternative process had to be used. Maskless projection lithography was a suitable choice during the electrode design phase since several designs had to be tested and optimized. The machine used during this process was developed during the PhD thesis of Ansgar Waldbaur [73]. This machine uses a digital micro-mirror device with an array of 1024x768 mirrors which can be individually controlled. Each individual mirror can be tilted to create either a dark or bright pixel. This allows the generation of dynamic masks. By using a demagnification optical system it is possible to achieve a line resolution of 1 µm. The demagnification also limits the size of a whole frame (comprising of 1024x768 pixels) to 2×3 mm^2, since the desired overall electrode size is 75×25 mm^2 it was necessary to "stich" approximately 300 frames next to each other to reach the desired size. This procedure was carried out to expose the AZ 1512 HS resist on a gold coated substrate. The fabrication of a mask using this approach produced a master electrode (or a gold mask) which was used to fabricate other electrodes using the setup described in 3.1.4.

For means of comparison a high resolution (25.400 dots per inch, equivalent to 10 µm line resolution) printed mask was also tested (see Figure 27 b).

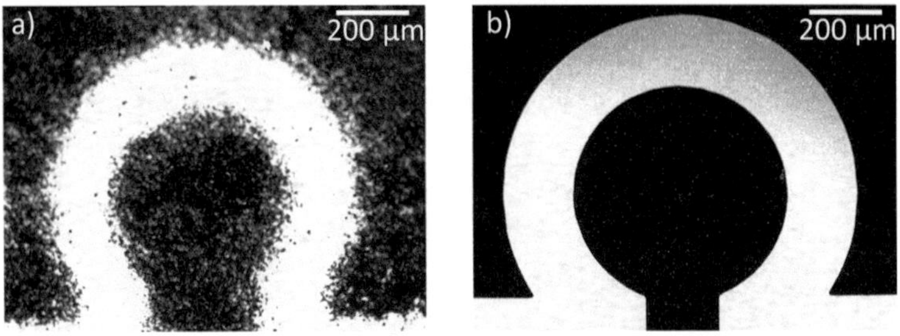

Figure 27 – Comparison of photolithography masks used during this work. a) Printed mask using a standard laser printer (100 µm line resolution). b) Printed mask from a professional high resolution (10 µm line resolution) plotter.

3.1.2 Gold sputtering

As explained in section 2.1.3 the electrodes have to be good electrical conductors, therefore metallic layers were chosen. Silver and copper are the best metallic conductors [74] but unfortunately they are cell-toxic [75, 76] and therefore unsuitable for this application. Gold, on the other hand is advantageous for numerous reasons. It is not cytotoxic [77] but has the highest oxidation potential (1,40; 0,80; 0,52 V against a standard hydrogen electrode for gold, silver and copper respectively [78]) of those three metals. This means that it is hardest to oxidize and therefore is extremely corrosion-resistant. Furthermore, it has the third best conductivity among metals. Therefore, gold was chosen for this work.

During the course of this work several substrate materials were tested including plastics (COC), ceramic and glass. They were chosen based on their chemical inertness to the required chemicals for the manufacturing process and their gold adhesion properties. Glass substrates require a thin (10 nm) adhesion layer of titanium or chromium, but due to the cell toxicity of chromium [79] titanium was chosen. Titanium thin film deposition as an adhesion layer for gold coating requires a sputtering machine which can deposit titanium and gold without breaking the vacuum because titanium is readily oxidized when in contact with air. A titanium dioxide layer does not promote a strong adhesion to gold. Titanium/gold glass substrates (see second step of Figure 23) were obtained from Axel Rosenhahn's group at the Institute of Functional Surfaces (IFG, KIT). All other gold depositions were done in-house.

COC substrates were cleaned in a 1:1 bidistilled water and isopropanol bath and dried with nitrogen. The cleaned substrates were placed in a DC sputtering machine of type Balziers SCD 040. The chamber was evacuated and argon was flushed into the recipient until a pressure of 0.05 mbar was reached. The plasma current was adjusted to 15 mA resulting in a gold sputter rate of 10 nm/min. The desired thickness was 100 nm (600 seconds).

3.1.3 Spin coating

Spin coating is a simple technique used to deposit thin films onto flat substrates. In this method the desired liquid or solubilized material is deposited in the middle of the substrate to be coated. This substrate is held by suction force on top of an axis by a vacuum pump. The axis rotates at high speeds (up to 10.000 rotations per minute, rpm), thus spreading homogeneously the material due to centrifugal forces. This process can be described using the approximation described in the equation below [80] in which h represents the film thickness, k a constant, C the concentration of the substance, μ the dynamic viscosity and ω the angular rotation velocity.

$$h = kC\left(\frac{\mu}{\omega^2}\right)^{\frac{1}{3}} \tag{9}$$

The spin coater used was WS-400-6NPP-LITE from Laurell Technologies Corp. 500 µl of AZ 1512 HS were pipetted on top of the substrate. The program chosen had two rotation steps: 500 rpm for 5 seconds and 3000 rpm for 20 seconds. The initial step was responsible for spreading the photoresist from the center to the edges of the substrate whilst the second step created a homogenous layer and removed the excess of photoresist.

After application of the photoresist the substrate is baked (post-exposure bake, 60 °C for 15 minutes) in order to remove the remaining solvents from the resist (part of which already evaporated during the rotation). After this step the coated substrate has to be kept in the dark to avoid undesired exposure.

3.1.4 UV light exposure

As explained in section 2.6, ultraviolet light induces chemical changes in the AZ photoresist used. Exposing the photoresist to UV light through a photomask will transfer the structure from the latter to the resist. The absorption range of AZ 1512 HS is from 310 to 440 nm peaking at 340 nm (see Figure 28). A mercury-arc lamp from Lumatec model Superlite 400 was used in this work. It has 10 different filter settings (position 0 to 9). The desired wavelength range is at filter position 3 (320 – 400 nm) with a rated average radiance emittance of 91.72 mW/mm^2 [63, 81].

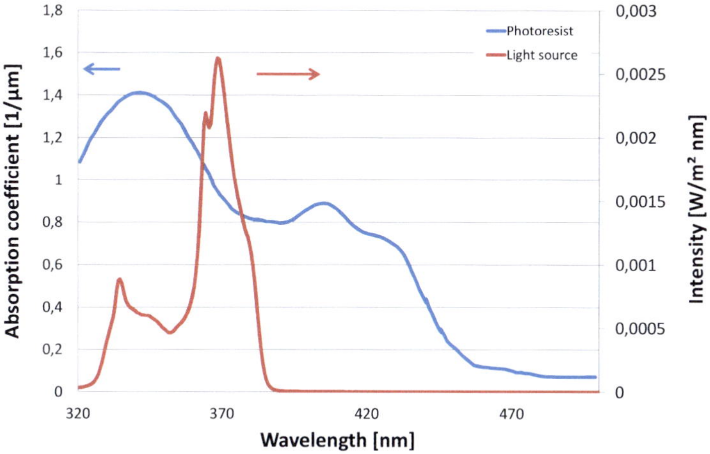

Figure 28 – Comparison between photoresist (AZ 1512 HS) absorption and light source emission (Lumatec Superlite 400 at filter position 3) as intensity over the wavelength.

The minimal feature size photolithography process can achieved is defined by (10) where k_1 and k_2 represent manufacturing coefficients, λ the wavelength and *NA* the numerical aperture of the lens system. According to the first equation, in order to further decrease the critical dimension the wavelength of the incident light should be decreased or the numerical aperture (*NA*) increased. However, there are certain constraints in minimizing the critical dimension, e.g., the depth of focus reduces when *NA* increases. If the resist thickness is higher than the depth of focus it will result in inhomogeneously exposed photoresist. Therefore, most of the efforts in reducing the critical dimension are focused on reducing the wavelength of the light used.

$$\text{critical dimension} = k_1 \frac{\lambda}{NA}$$
$$\text{depth of focus} = k_2 \frac{\lambda}{NA^2} \qquad (10)$$

Several exposure tests had to be conducted to determine the ideal exposure duration for a spot size of 10 cm diameter (see Figure 29), this value turned out to be 60 seconds at the maximum lamp power intensity.

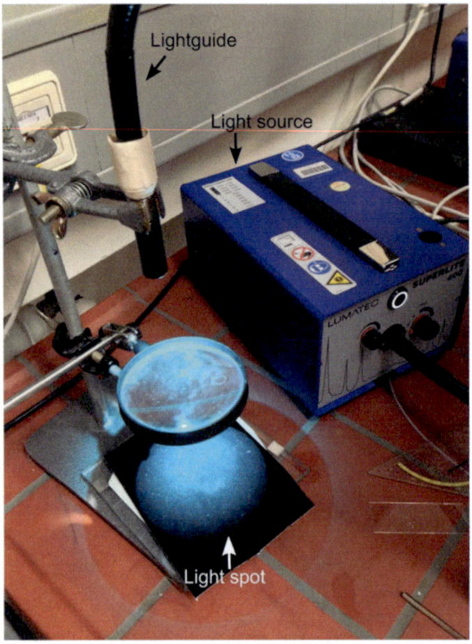

Figure 29 – Setup used for UV exposure. The light source used is from Lumatec model Superlite 400. The light guide redirects the light beam generated by the mercury lamp onto the converging lens. A light spot of approximately 10 cm diameter with suitable collimation is obtained.

3.1.5 Resist Developing

As explained in section 2.6, exposed AZ photoresists become soluble in basic solutions. The photoresist manufacturer sells a developer solution called AZ 351B which contains a concentrated NaOH and $Ba(OH)_2$ solution and has to be diluted 1:4 in water before being used. The exposed substrate was put in a bath containing this solution for 2 minutes. Afterwards the substrate is purged thoroughly with bidistilled water and dried with nitrogen in order to stop the developing. The exposed areas start to dissolve almost instantaneously. After the conclusion of this step the photomask structure can be clearly seen on the photoresist (see sixth step of Figure 23).

3.1.6 Gold etching

The structured photoresist on top of the gold layer serves as a protective mask to the gold etchant solution. The areas which are covered by the photoresist will not be readily removed by the etchant. The chosen etchant was iodine/potassium iodine (K/KI_2) a cheap and very efficient solution to selectively etch gold. The oxidation reaction is as follows [82]:

$$2Au_{(s)} + I_{2(aq)} \rightarrow 2AuI_{(aq)} \tag{11}$$

Potassium iodine is used to increase the solubility of the gold-iodine complex in water thus enabling the same solution to be used longer whilst increasing the etch rate. At a ratio between KI, I_2 of 4 g to 1 g in 40 ml of water the etch rate is around 1 µm/min at room temperature [83].

The coated substrate is put in a bath of the etchant solution but due to the reuse of this solution it is not reliable to use a predetermined etch duration. The etching is controlled visually and is stopped by rinsing the substrate with water and drying with nitrogen (see Figure 23).

The remaining photoresist on top of the gold structures was then removed in a bath of acetone and ultrasonicated for 2 minutes. In order to avoid acetone smears the electrode is washed with isopropanol followed by bidistilled water and dried in nitrogen flow (see Figure 23).

3.1.7 Titanium etching

As mentioned in section 3.1.2 glass substrates require an adhesion layer for gold deposition. In this work a titanium layer was used. As a consequence, the areas from which gold was etched still contain the underlying gray titanium adhesion layer. Titanium is not an exceptional conductor but nonetheless its conductivity is 1/6 of the conductivity of gold (and around 200 µΩ/cm for 10 nm thin films, [84]) which is high enough to shortcut all electrodes. It is thus necessary to etch this layer as well. Titanium can only be etched by very strong oxidizing agents [85], e.g., hydrofluoric acid and piranha solution, a 1:1 volume solution of H_2SO_4 and H_2O_2. Piranha solution was chosen in the course of this work due to its easier handling. A total volume of 120 ml of piranha solution was poured slowly into a glass vial into which the substrates were subsequently immersed. After 5 minutes in the solution the substrates were removed and rinsed copiously with water and dried under nitrogen flow. After etching of the titanium layer the unprotected substrate areas became completely transparent (see Figure 23).

3.1.8 Implementation of reference electrode

Some experiments throughout this work required the implementation of a three-electrode measurement setup (see Section 2.1.4 for details) due to the need of accurately determining the potential applied to the electrodes. This was implemented after the electrode was fabricated by manually applying a drop of Ag/AgCl paste to one of the gold electrode pairs (see left electrode in Figure 30). Afterwards the substrate was baked at 60 °C for 20 minutes in order to remove the solvent from the paste.

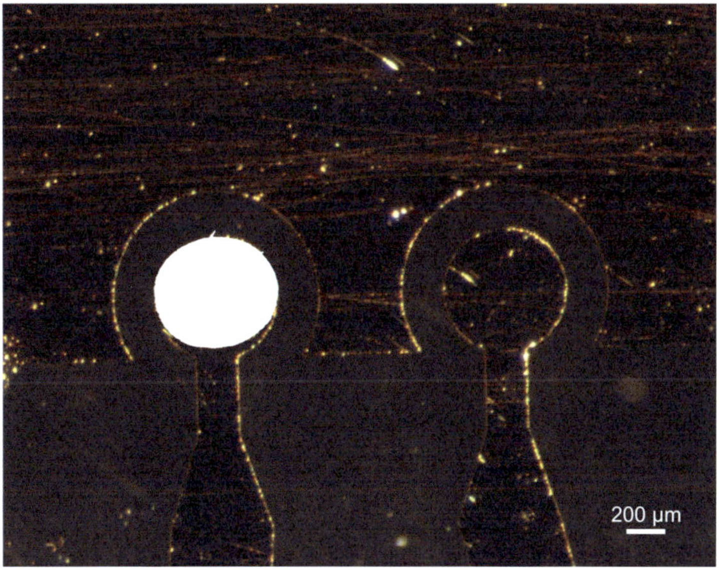

Figure 30 – Microscopy picture of electrode pair. Ag/AgCl paste was applied to the left electrode and is seen as a white circle.

3.2 Measurement electronics

The impedance spectrum analyzer (see Figure 38) used during this work was designed, developed and tested in-house prior to this work by Kai Sachsenheimer. The schematic block diagram can be seen in Figure 31. The digital values of a sinusoid are read from a lookup table in the microprocessor (8 bit ATmega32) and then low pass filtered and amplified (signal $A_0 sin(\omega t)$). This signal is applied between working and ground electrodes, resulting in a current which is attenuated and phase shifted yielding the current $I_0 sin(\omega t + \emptyset)$. The current is then converted into a voltage and multiplied by the sinus wave (input signal) and by a cosine wave generated by the microcontroller in order to separate the real from the imaginary part of the signal. After low pass filtering with a cut frequency lower than the original input signal frefrequency both signals are analog-to-digital converted and the real and the imaginary part of the complex impedance is obtained. This procedure is repeated 100 times for each frequency point and averaged. The measurement electronics is capable of measuring impedance spectra from 10 Hz up to 100 kHz. The amplitude of the signals applied is 20 mV peak-to-peak. This low amplitude does not disturb or damage the molecular structure of the sample which may contain proteins, antibodies and enzymes. For the amperometric measurements performed during the biofilm experiments an amperometric module was developed. This module is responsible for measuring the metabolic current produced by a biofilm. The current is measured by monitoring the voltage drop across a 1 MΩ resistor. The voltage is converted by a 16-bit analog-to-digital converter. The measurement electronics can measure currents down to 100 pA. Nine coaxial cables connect the eight electrodes and ground to the measurement electronics.

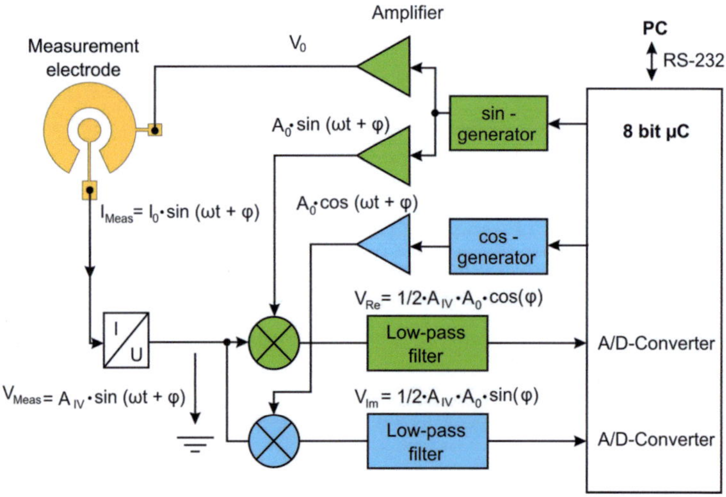

Figure 31 – Block diagram of the custom-made impedance spectrum analyzer.

Signals are transferred to the measurement computer via serial port (RS-232). The microprocessor will relay this information together with the frequency to the computer. The software communicating with the electronics is a terminal program (Terminal V1.9b) which reads the input from the electronics in regular intervals and saves it in a text file.

3.3 Connection strategies

Several different connection techniques were used to connect the planar gold electrodes to the measurement electronics. Soldering cables directly to the planar electrodes damaged them and did not provide a suitable electrical connection. Initially metallic pins were glued to the substrate and by means of a conductive silver paste electrically connected to the planar electrodes (see Figure 32 a). These metallic pins can be inserted in their respective female connectors which are soldered to the cables attached to the electronics. This was complex and time consuming (included several baking steps for the silver paste). Therefore this connection strategy was replaced by a custom made conductive silicon pad which contained silver particles. These pads were perforated by wires soldered to metallic pins identical to those used in the previous strategy, therefore electrically connecting each other. The metallic pins could then be connected to the measurement electronics (see Figure 32 b). By pressing the silicon pads against the planar gold surface electrical contact was established. However, after several connections and disconnections procedures the silicone pads showed signs of wear and damage and had to be replaced. Therefore a third (and preferred) concept was developed, using spring-loaded connector slots as found in computers mother boards (see Figure 32 c). This approach is convenient to implement and allows fastest connection time.

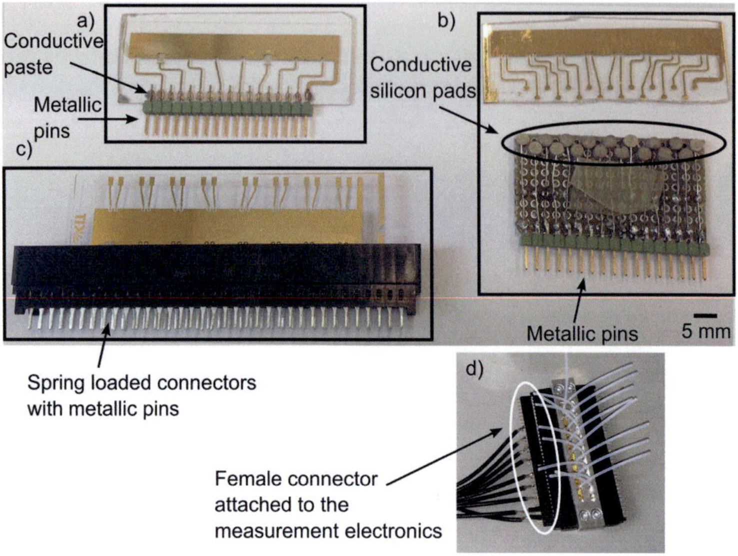

Figure 32 – Comparison between the three electrode connection strategies developed during this work. a) Metallic pins glued to the substrate and connected to the planar gold electrodes using a conductive silver paste. b) Conductive silicone pads containing silver particles are pressed against planar gold electrodes pads. c) Electrode substrate after insertion into spring-loaded slot connectors. d) A female connector soldered to cables attached to the measurement electronics is responsible for connecting the electronics to the metal pins seen on all strategies.

3.4 Microfluidic flow cell

The planar electrodes need to be individually probed and therefore a microfluidic channel structures, a so-called as flow cell, is required. In this work the PDMS flow cell was directly cast against an epoxy resin (Accura® 60) which was structured using stereo lithography (by PROFORM AG). The casting structure was designed in 3D computer aided design (CAD) software (see Figure 33). The PDMS used was Wacker

Elastosil 601. According to the manufacturer's instructions 9 parts by weight of component A are mixed to 1 part by weight of component B of the PDMS prepolymer. After degassing the mixture for 5 minutes in partial vacuum, the PDMS prepolymer was poured into the cast and tempered for one hour at 60 °C. In order to reduce analyte diffusion into the material and increase the chemical resistance (as required, e.g., for the azo coupling) the flow was coated with 100 nm layer of parylene C by means of chemical vapor deposition.

Several different channel design parameters (height, length, width, number of channels and distance between channels), were tested throughout the course of this work and resulted in three main layouts with one, eight and sixteen channels (see Figure 33). All channels tested were 100 µm deep and 1 mm wide. The eight and sixteen electrode flow cells have channels of 4 mm length whilst the one channel and two-channel flow cells have channel lengths of 53 and 18 mm, respectively.

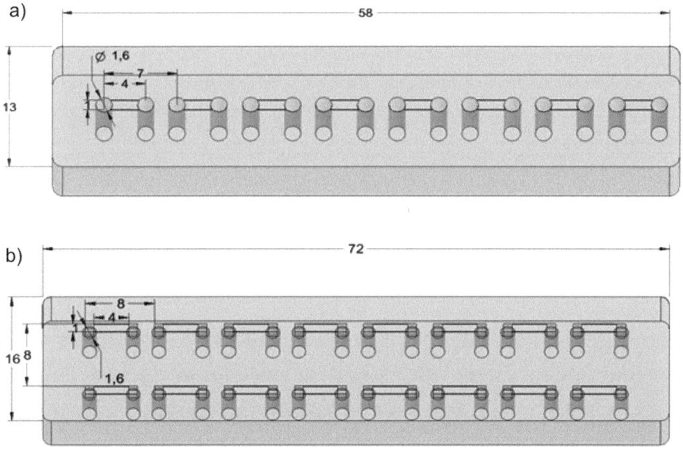

Figure 33 – 3D CAD drawings of two different flow cells with relevant dimension indications. All channels are 100 µm deep. a) Eight-channel flow cell used predominantly for the plastic substrates (6 cm long). b) Sixteen-channel flow cell used predominantly for the glass substrates (7,2 cm long).

The flow cell is pressed onto the substrate by means of an aluminum housing which clamped the parts by means of four M4 screws (see Figure 34). During this work four different housing designs were used and evaluated according to the current connection strategy.

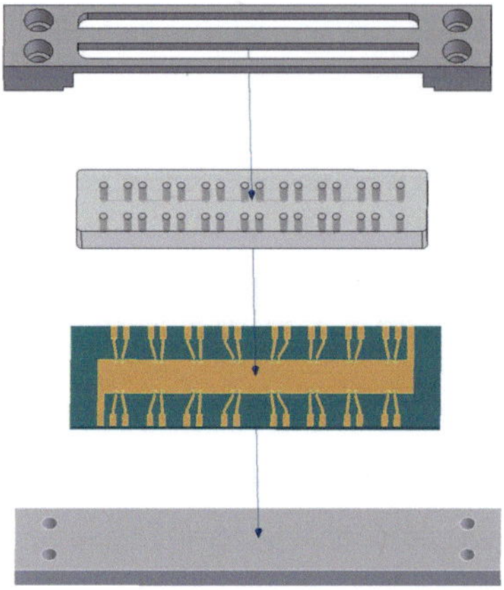

Figure 34 – Explosion view of the latest measurement cell setup. PDMS flow cell and electrode substrate are clamped together by means of two aluminum plates and four M4 screws. The flow cell's channels are aligned with the electrodes forming a channel on top of them.

The first housing design (see Figure 35 a) was bulky and was used initially only with glued contact pins. Due to the disadvantages of this connection method (see Section 3.3) it was replaced by the conductive silver pads which could also be used with this first generation housing design. The second generation (see Figure 35 b) was trimmed on one side in order to allow usage of spring loaded connectors while also providing backward compatibility with the first two connection strategies. The change from plastic substrate to glass also required changes in the flow cell (as the substrate size was different) and consequently also in the housing design resulting in the third housing generation. Due to the simplicity and ease of use the spring loaded connector became the standard electrode connector and in order to increase the number of electrodes per substrate the flow cell was further adapted to host sixteen electrodes in the third generation (see Figure 35 c). The fourth generation (see Figure 35 d) became the standard housing and was implemented because it provided better mechanical stability due to its thicker metal frame.

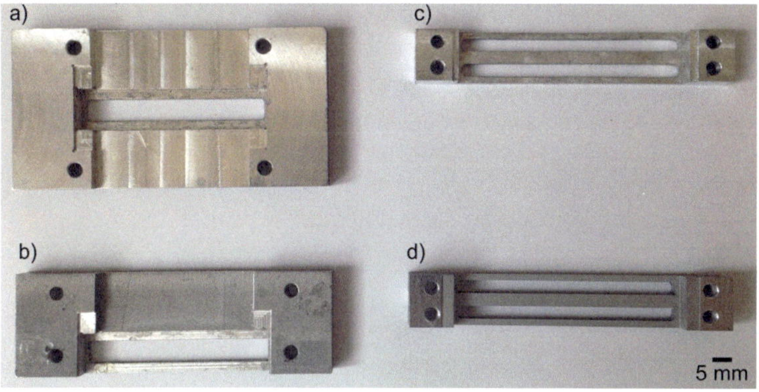

Figure 35 – The four different flow cell housing organized chronologically. a) First housing for 60 mm length flow cell using glued contact pins or conductive silicon pads. b) Second housing generation designed for glued contact pins, conductive silicone pads and spring loaded connectors. c) Third generation housing generation suitable for use of spring loaded connectors. d) Forth housing generation with increased mechanical stability due to increased thickness of the metal frame.

3.5 Electrode design

Experiments were conducted in order to assess the influence of the electrode design upon the initial impedance values, sensor sensitivity, drift behavior and saturation threshold. Four different electrode designs were conceived and are depicted in Figure 36. An interdigitated electrode with 200 µm pitch and 100 µm wide fingers (see Figure 36 a) and three round electrodes with 600, 400 and 200 µm in diameter on a constant 1 mm circular cutout (see Figure 36 b, c and d respectively) are shown.

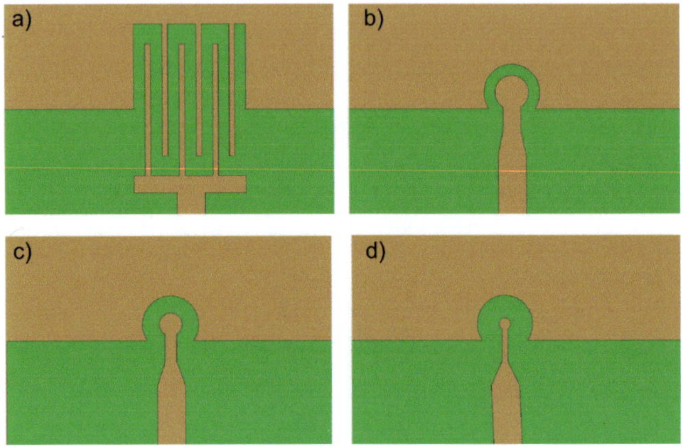

Figure 36 – 3D CAD view of the four different electrode designs tested.
a) Interdigitated electrode with 200 µm pitch and 100 µm wide fingers. b) Round electrode with a 600 µm diameter and a 200 µm gap. c) Round electrode with a 400 µm diameter and a 300 µm gap. d) Round electrode with a 200 µm diameter and a 400 µm gap.

The experiments firstly evaluated the initial impedance and drift behavior of each electrode by conducting two measurements in a five minute interval. Afterwards the electrodes were probed with 0.01 µg/ml and 1 µg/ml of BSA in PBS (300 µl at 50µl/min), respectively, in order to assess sensitivity and saturation threshold of the electrodes. All measurements were carried out in electrolyte solution (25 mM of $K_3[Fe(CN)_6]$, 25 mM of $K_4[Fe(CN)_6] \cdot 3H_2O$ and 100 mM of KCl) as opposed to performing the binding event and measurement in the same solution (e.g., analyte solved in electrolyte solution). The measurements took place after purging the electrode with the electrolyte solution for one minute at 50 µl/min. This was done for three reasons; to purge unbound BSA molecules from the surface, to stop the binding process after 6 minutes of sample probing (300 µl at 50 µl/min) and to provide a reproducible solution conductivity for each measurement (independent from the analyte concentration). All these measures were taken to increase result reproducibility. Exemplary results of these measurements can be seen in Figure 37. As expected, the electrodes with smaller gaps showed smaller overall impedances values (see Figure 37).

The impedance is shown in a Nyquist diagram where the vertical axis represents the module of the imaginary impedance component (reactance) and the horizontal axis the real part (resistance). The interdigitated electrode showed the smallest initial impedance values and highest sensitivity but also the highest drift and the lowest saturation threshold (see Figure 37 a). The electrode with 600 µm in diameter showed the second lowest initial impedance, comparable sensor sensitivity, a small drift and a high saturation threshold (see Figure 37 b). The electrode with 400 µm in diameter showed higher initial impedance, low sensitivity, small drift and high saturation threshold (see Figure 37 c). The impedance of electrodes with 200 µm in diameter electrodes (Figure 36 d) was much larger than all of the other electrodes and is not shown. A larger gap between the round electrode and the working electrode results in high impedance values in lower conductivity media (e.g., electrolyte solution in comparison to gold). Based on this data, the largest round electrode (Figure 36 b) design was chosen as the standard electrode format for this work.

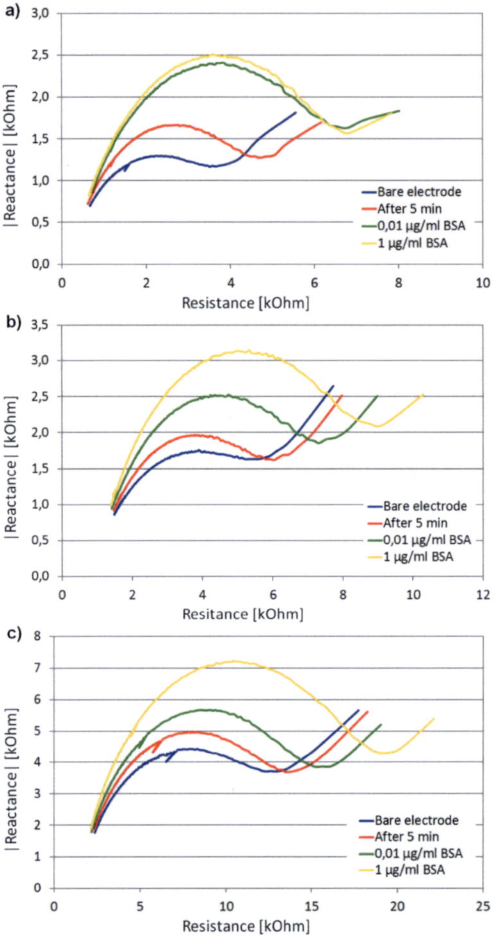

Figure 37 – Nyquist diagrams for the nonspecific adsorption assay of bovine serum albumin (BSA) in PBS on the bare gold surface of different electrode geometries from Figure 36 a, b and c, respectively. 300 µl BSA samples in PBS were injected at 50 µl/min into the channel for 6 minutes. After one minute of purging the channel with electrolyte solution at 50 µl/min the pump was turned off and the measurement was carried out.

3.6 Measurement setup

The measurement setup (see Figure 38) contains the measurement cell which consists of the electrode substrate with the flow cell clamped to it by the flow cell housing. The measurement cell is connected to the measurement electronics and its power supply. The measurement electronics is connected to the computer via a RS232 connection. The liquid is pumped via a peristaltic pump from a reservoir through the measurement cell into a waste reservoir. Two different EIS methods could be used with this system, faradaic and non-faradaic (see Section 2.1.7). Faradaic EIS systems allow the current to flow through the electrode-electrolyte interface, thus providing additional kinetic information about the biochemical binding reaction taking place on the surface of the electrode. Therefore, a faradaic method was chosen which in turn required the use of an electrolyte solution during the EIS measurements. This solution consists of 25 mM of $K_3[Fe(CN)_6]$, 25 mM of $K_4[Fe(CN)_6] \cdot 3H_2O$ and 100 mM of KCl. As previously mentioned (see Section 3.5), the measurements were carried out in electrolyte solution in order to increase the reproducibility of the measurements.

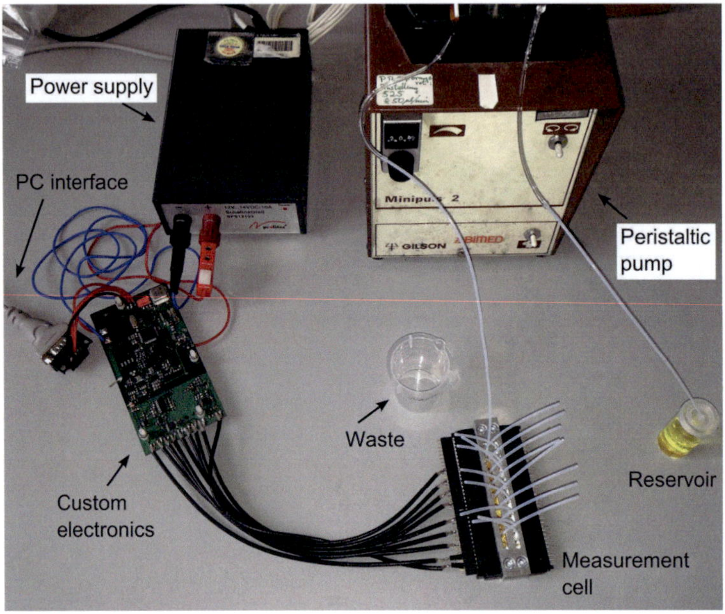

Figure 38 – Measurement setup consisting of a peristaltic pump, measurement electronics with power supply, measurement cell (flow cell housing, flow cell, electrode substrate), electrolyte reservoir and waste. Tygon pump tubing with 0.76 and 2.46 mm inner and outer diameter, respectively, was used. All other tubings were polytetrafluoroethylene (PTFE) tubing with 0.8 and 1.6 mm inner and outer diameter, respectively.

3.7 Biofilm detection

Another application field which was investigated during the course of this work was the monitoring of biofilm (see Section 2.4) using EIS and amperometry. The detection of biofilm growth by means of EIS is, compared to an affinity assay, more straightforward as it does not require a specific surface modification of the electrodes (see Section 2.5.2). As the bacterial film colonizes the electrode surface it gradually hinders the electron transfer between the electrodes resulting in an impedance increase (see Figure 39).

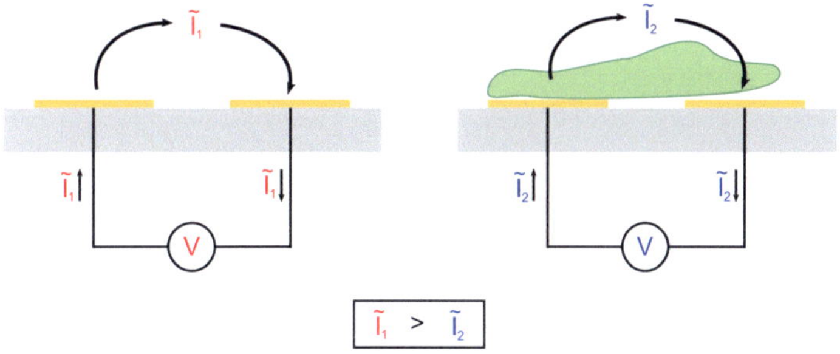

Figure 39 – Diagram of two electrode systems during two different biofilm colonization stages. a) The current between the electrodes flows unhindered in the absence of biofilm. b) The formation of a biofilm on top of the electrode hinders the charge transfer between the electrodes, thus increasing the measured impedance.

Monitoring the growth of the biofilm is just one parameter relevant to understanding the behavior of bacterial colonies. Information about the respiratory activity of the biofilm provides crucial information for the development of treatment procedures for the disintegration or deactivation of these bacterial colonies. The respiratory processes of bacteria can be assessed by monitoring the resulting extracellular and membrane associated charge transfer [86]. In order to do so an additional amperometric module was developed to monitor the extracellular current resulting from these respiratory processes. This two-pronged approach (see Figure 40) allows for real-time monitoring of the biofilm film growth and respiratory activity.

3.7.1 Setup

The setup for this combined measurement approach consists of a flow cell which contains two independent channels, each housing three amperometric electrodes and four planar EIS gold electrodes, a custom made electronics and a flow cell housing. One microfluidic channel (reference) was not probed with bacteria solution, the other microfluidic channel (measurement) was probed in order to allow biofilm growth in this channel. Having a reference and a measurement channel allows reducing signal drifts caused by ambient temperature variations by subtraction of the two signals over time. The measurements were carried out using gram-negative, rod-shaped, and opportunistic pathogen *Pseudomonas aeruginosa* type UCBPP-PA14. The biofilm was monitored during seeding, growth and treatment with a biocide (sodium azide). The results obtained by means of EIS and amperometry were confirmed by fluorescence microscopy after live/dead staining of the bacteria.

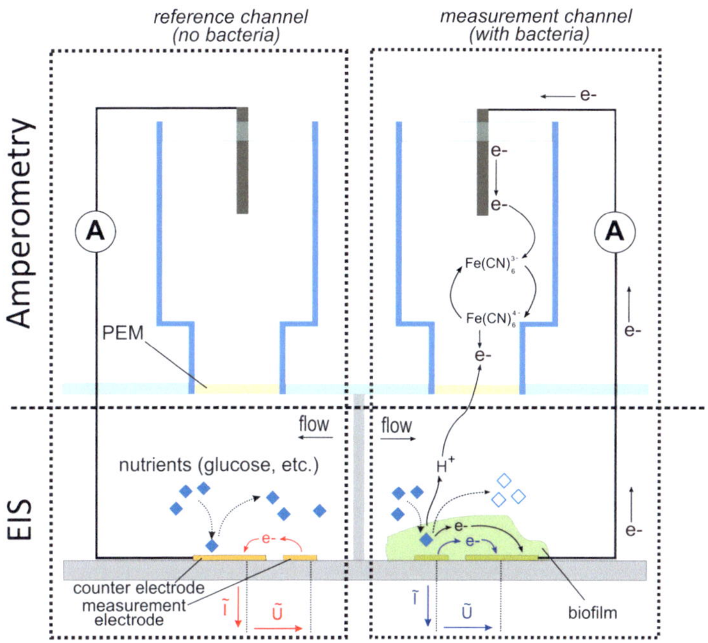

Figure 40 – Measurement setup schematic. Two independent fluidic channels are shown: reference channel (left) and measurement channel (right). During experiments a biofilm was seeded only in the measurement channel. Each channel features four electrodes for EIS and three for amperometry measurements. This setup enables the measurement of the biofilm growth (EIS electrodes) and its respiratory activity (amperometry electrodes). Published in [87].

EIS can determine the presence of a biofilm on top of the electrodes and asses the film thickness but it gives no information on whether or not the biofilm is alive or dead. This is problematic because many biofilm colonies do not detach from the surface after being killed which results in dead biomass being deposited potentially serving as feeding ground for further biofilm development. Bacteria whose respiratory processes are linked with membrane-associated enzymes or exoenzymatic activities (oxidases, reductases) enable the monitoring of their respirato-respiratory activity by measuring the created extracellular and membrane-associated charge transports, i.e., currents. Amperometric methods enable the monitoring of a current over time and are therefore a suitable measurement technique for respiratory activity assessment. The combination of EIS and amperometry allows gaining insight into the dynamics of biofilm behavior when treated with different biocides.

3.7.2 Measurements

Measurements were carried out with the aim of monitoring the biofilm behavior during repeated biocide (sodium azide) probing (see Figure 41). Firstly, a bacterial suspension solution of *Pseudomonas aeruginosa* was seeded in the measurement channel during the biofilm seeding phase (interval A in Figure 41) for 3 hours. As can be seen there is a clear increase in the measured impedance which correlates with the biofilm seeding on the surface. Furthermore, a significant increase in the amperometric signal can be observed which results from the high concentration and thus high respiratory activity of the bacteria in close contact with the electrode. After seeding the biofilm was grown for 72 hours by probing brain heart infusion (BHI) broth through the microfluidic system. As can be seen both EIS and amperometric signal increase over time (interval between A and B in Figure 41). After the growth phase the measurement channel is probed with a solution of 1 m% sodium azide for 4 hours (interval B in Figure 41). As can be seen the impedance reaches a plateau whereas a clear decline in the current curve is shown. This indicates that the biofilm significantly reduces its respiratory activity which indicates that the bacteria are dying. However, based on the non-declining EIS data it can be established that the overall biomass is not reduced. Directly after this biocide probing a second growth interval (between B and C in Figure 41) of 40 hours was applied during which the biofilm regrew as clearly indicated by the rise of the EIS signal (increase in biomass) and the increase in the amperometric signal (increase in respiratory activity). After the second growth phase a second biocide probing of sodium azide (1 m%) for 6 hours took place (interval C in Figure 41) analogous to the previous biocide injection. Again the EIS signal shows a plateau whereas the amperometric signal decreases. As discussed, the bacteria are largely affected by the biocide but the dead biomass does not dissolve from the substrate. After this second biocide probing a final growth phase of 24 hours took place (between C and D in Figure 41) which again is marked by a slight increase in both monitored parameters. A final biocide (sodium azide 1 m%) injection phase after 24 hours of growth took place for the remaining 20 hours of the experiment (interval D in Figure 41) which shows a drastic current decrease and almost unaltered impedance values again indicating a strong decline of biofilm respiratory activity and no biofilm disintegration.

This experiment efficiently demonstrated the suitability of this combined electrochemical measurement system for characterizing mass and activity of a biofilm on a technical substrate. The use of this system demonstrated that sodium azide as a treatment agent is inefficient due to the fact that although it kills the bacteria of the biofilm it is unable to dissolve the biofilm. Quickly after the end of biocide probing the biofilm recovers and accumulates more biomass over time. This trend could neither be prevented nor reverted by sodium azide.

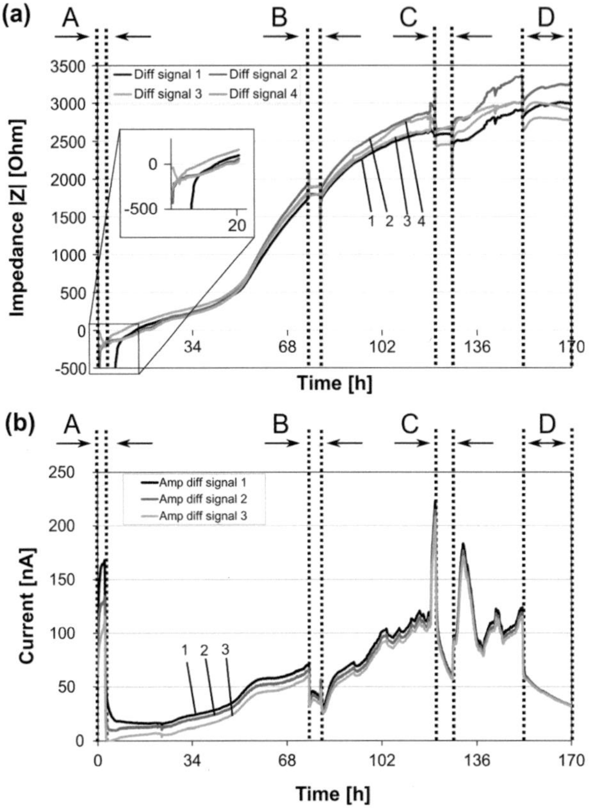

Figure 41 – Experimental results for combined electrochemical impedance spectroscopy (a) and amperometric biofilm growth characterization (b) during three (B, C and D) injection periods of sodium azide (1 m%). A – Initial injection of *Pseudomonas aeruginosa* suspension (approximately 10^8 colony forming units per ml) during 3 hours. Three growth intervals took place during this measurement, between A and B, B and C and C and D; for 72, 40 and 24 hours respectively. A constant flow of BHI was probed across both channels during these intervals. B, C and D represent three sodium azide (1 m%) injections for 4, 6 and 20 hours, respectively. Published in [87].

4 Development of low impedance surface modifications

Surface modification encompasses a range of different methods to modify a surface in order to provide functional groups for the immobilization of analyte specific molecules whilst simultaneously suppressing the nonspecific adsorption of unwanted molecules (see Section 2.5 for details). This section highlights the surface modification techniques implemented in the course of this work and the experimental results obtained from electrode systems modified using the respective surface modification technique. Advantages and disadvantages of each method will be indicated. The method of choice is a process based on photobleaching which has been developed and adapted to biosensing during the course of this work. This method is not known from literature and this work is the first to report usage of this versatile technique.

4.1 Adsorption experiments on unmodified electrodes

The first experiments carried out were based on protein adsorption on unmodified gold electrodes. These experiments were carried out in order to assess the sensor's sensitivity. Different concentrations of a protein (bovine serum albumin, BSA) in PBS ranging from 1 ng/ml to 1 µg/ml were probed across a bare gold electrode (see Figure **42**). The initial measurement (Figure **42**, blue curve) establishes the initial impedance of the sensor. The frequency range chosen was from 10 Hz to 5.000 Hz (and later extended to 50.000 Hz). The flow rate was set to 50 µl/min. In order to perform faradaic EIS measurements (see Section2.1.7) an electrolyte solution (25 mM of $K_3[Fe(CN)_6]$, 25 mM of $K_4[Fe(CN)_6] \cdot 3H_2O$ and 100 mM of KCl) was probed through the microfluidic channel for one minute at 50 µl/min, the pump was then turned off and a measurement was carried out assessing the changed surface impedance. Performing the EIS measurements in pure electrolyte solution (as opposed to analyte solved directly in the electrolyte solution) has many advantages. Firstly, the purging step removes unbound molecules from the electrode surface, thus minimizing the effects of unbound molecules onto the measurement. Secondly, it limits the binding process to the sample probing interval (300 µl at 50 µl/min) and finally, it provide a reproducible solution conductivity for each measurement (independent from the analyte concentration). All these measures were taken to increase result reproducibility.

As can be seen the impedance increases gradually after each probing with increasing concentration.

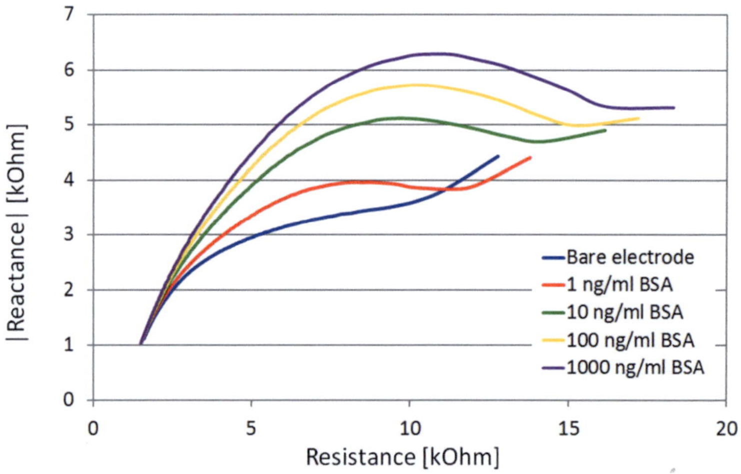

Figure 42 – Nyquist diagram of the nonspecific adsorption of bovine serum albumin (BSA) on bare gold surfaces. All measurements were carried out in electrolyte solution in stationary flow and from 10 Hz to 5000 Hz. All EIS measurements were preceded by a one minute electrolyte solution purge.

This experiment showed that the sensor is able to clearly detect a protein of 1 ng/ml concentration. This sensitivity threshold gives a rough estimate for further experiments. This detection limit is in the same range as found for reference methods based, e.g., on gel electrophoresis (for instance, sodium dodecyl sulfate polyacrylamide gel electrophoresis – SDS-PAGE). However, the sensor is not specific. Using unmodified electrodes, the sensor would react to any protein in an identical fashion thus rendering discrimination of analytes impossible. As stated, in order to transform this system into a biosensor it is necessary to modify the electrode surface such that it can fulfill the requirements described in 2.5.

4.2 Cleaning procedures

Before applying surface modifications the sensor surface must be sufficiently cleaned from impurities. Several cleaning methods are known from literature and have been evaluated. It is of paramount importance to have the cleanest gold surface possible before starting the surface modification procedure as contaminants may hinder the formation of the surface modification and may increase the initial sensor impedance considerably. This not only decreases the sensor's sensitivity but usually increases electrode-to-electrode variations significantly (see Figure 43).

Several methods listed in literature [88] were tested throughout this work, e.g., plasma cleaning, cyclic voltammetry (CV) in sulfuric acid, CV in KOH and immersion in piranha solution. Plasma cleaning is a very simple procedure but the method was found to yield electrodes with high variations in surface impedance. Cyclic voltammetry in a 10 mM sulfuric acid aqueous solution was tested as an alternative method. However, before a stable cyclic voltammogram could be reached (around the second cycle from -400 to 1400 mV versus Ag/AgCl) the electrodes showed signs of strong corrosion. Alternatively cyclic voltammetry (-200 to 1200 mV versus Ag/AgCl) in a 50 mM solution of KOH was carried out and yielded good cleaning results with small electrode-to-electrode variations. The disadvantages of this method are two-fold. Firstly, it is time-consuming (around one hour) as the electrodes have to be washed thoroughly afterwards. Secondly, it requires the application of Ag/AgCl paste to the reference electrodes which in turn adds 30 minutes to the electrode fabrication procedure due to the necessity of baking the paste (see Section 3.1.8). Immersing the whole substrate in piranha solution was found to be the best method as it is fast (30 minutes for a batch of up to 10 substrates) and produced substrates with minor electrode-to-electrode variations (see Figure 44).

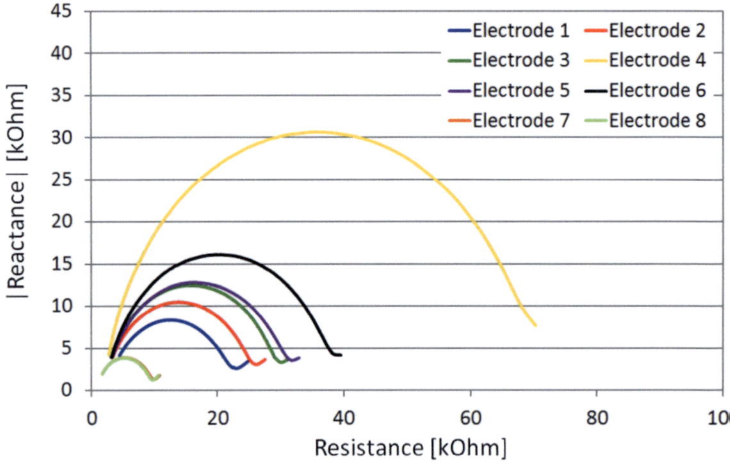

Figure 43 – Nyquist diagram of eight gold electrodes from the same substrate functionalized with a substance which was used for creating unsaturated surfaces used for photobleaching (see Section 4.6), i.e., allyl mercaptan. All EIS measurements were preceded by a one minute electrolyte solution purge and carried out in electrolyte solution.

The EIS frequency range was extended to 50.000 Hz in order to provide more information about the electrode at high frequencies. It is important to note that the last step of the electrode manufacturing (see Section 3.1.7) also consists of immersing the substrate in piranha solution. However, the electrodes were usually not used directly after manufacturing but rather stored for varying amounts of time before being used. This rendered this additional cleaning step prior to surface modification necessary.

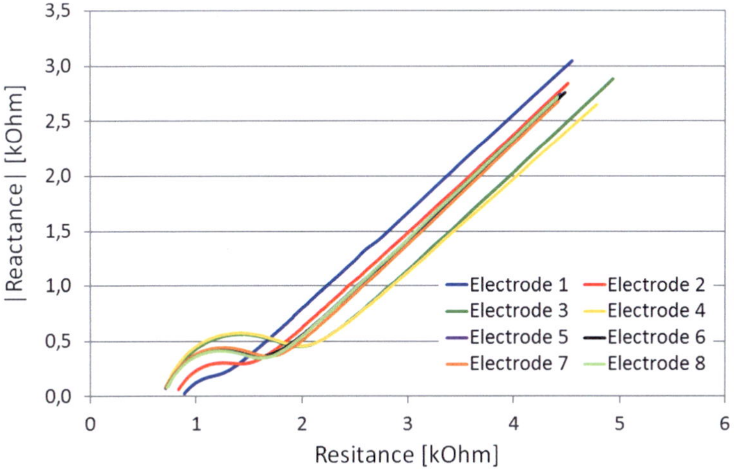

Figure 44 – Nyquist diagram of eight gold electrodes from the same substrate functionalized with allyl mercaptan (see Section 4.6), which were cleaned with piranha solution before the application of allyl mercaptan. All EIS measurements were preceded by a one minute electrolyte solution purge and carried out in electrolyte solution. Very low initial impedance values can be obtained and the Warburg impedance dominated region is clearly visible. Electrodes show good electrode-to-electrode reproducibility.

Measuring the contact angle on glass and gold surfaces is another method to assess the cleanness of the surfaces. After immersing glass in piranha solution it becomes superhydrophilic as piranha solution readily oxidizes the surface creating high concentrations of surface-bound silanols. According to literature, ordinary gold has a contact angle of 60° to 65° [89]. Contact angle measurements after the cleaning with piranha solution resulted in values around 61°.

4.3 Self-assembled monolayer (SAM)

One of the most common methods found in literature for surface modification is usage of substances that form SAMs. This technique is appealing due to its simplicity and ease of use (see Section 2.5.5 for details). Several SAM-forming compounds, i.e., 3-Mercaptopropionic acid (3-MPA), 6-Mercaptohexanoic acid (6-MHA), 8-Mercapto-octanoic acid (8-MOA), 11-Mercaptoundecanoic acid (11-MUA) and 4-Mercaptobenzoic acid (4-MBA) were investigated during the course of this work (see Table 2). A 2 mM solution of the SAM-forming substance in ethanol was pipetted onto the cleaned gold electrode and incubated overnight. Afterwards, the electrode was washed with ethanol and bidistilled water and dried under nitrogen. All substances underwent the same initial characterization using EIS. After SAM formation the electrodes were probed three times with a highly concentrated (1 mg/ml) BSA in PBS (300 µl at 50 µl/min) solution to assess the presence of holes or imperfections in the formed SAMs. As BSA readily adsorbs and adheres to gold surfaces measurable increases in the surface impedance will indicate imperfections on such layers (see Section 4.1 and Figure 42 which discuss an adsorption experiment).

Experiments with 3-MPA showed a very low initial combined impedance (magnitude of the resistance and reactance) of 61 kΩ which was expected given the fact that this is a short molecule and not expected to introduce highly insulating layers. However, BSA adsorption was found to be very high, probably due to an insufficiently dense SAM which may have resulted from low Van der Waals forces displayed by such a short carbon chain and thus low self-ordering effects in the SAM (see Figure 45).

	3-Mercapto propionic acid	6-Mercapto hexanoic acid	8-Mercapto octanoic acid	11-Mercapto undecanoic acid	4-Mercapto benzoic acid
Chemical structure	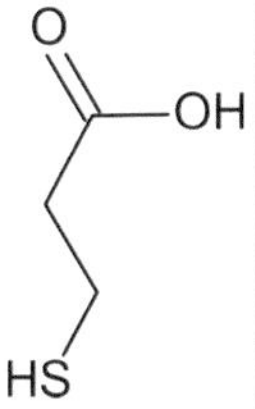	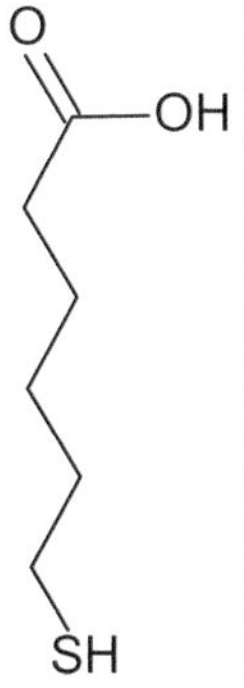	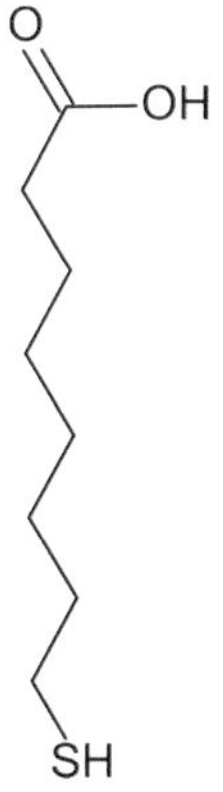	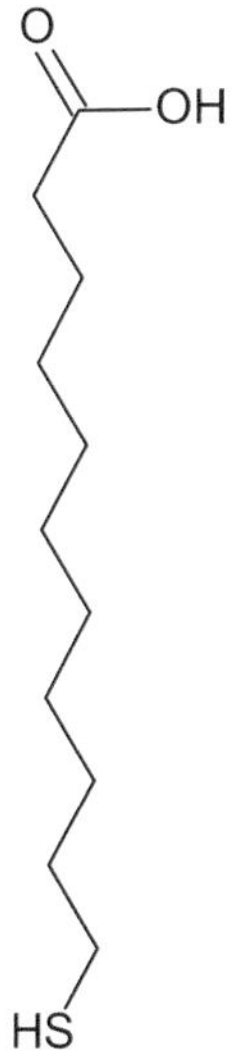	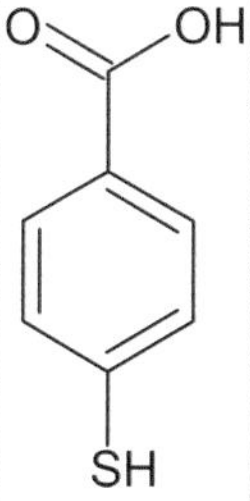
|Z| at 10 Hz	61 kΩ	430 kΩ	769 kΩ	1600 kΩ	220 kΩ

Table 2 – Overview of the SAM-forming substances tested during this work. The values present in the bottom part of the table represent the magnitude of the impedance at 10 Hz for each tested substance.

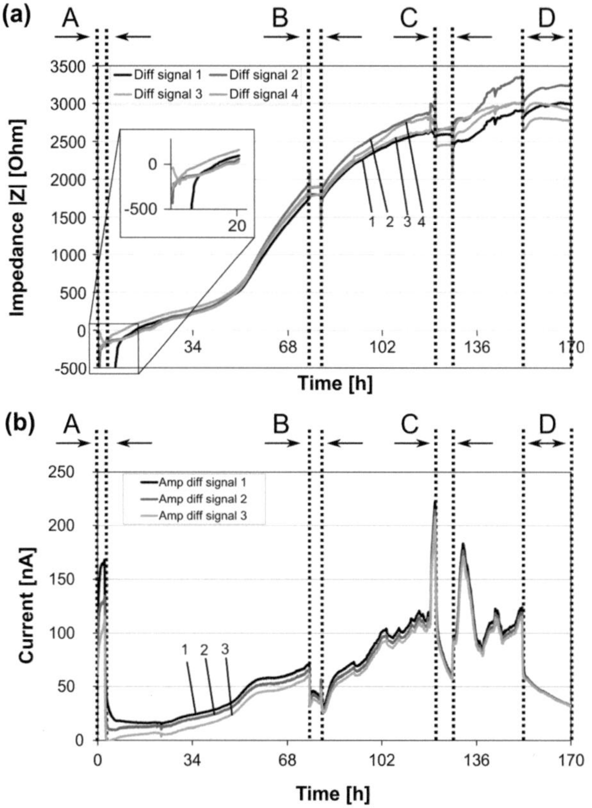

Figure 41 – Experimental results for combined electrochemical impedance spectroscopy (a) and amperometric biofilm growth characterization (b) during three (B, C and D) injection periods of sodium azide (1 m%). A – Initial injection of *Pseudomonas aeruginosa* suspension (approximately 10^8 colony forming units per ml) during 3 hours. Three growth intervals took place during this measurement, between A and B, B and C and C and D; for 72, 40 and 24 hours respectively. A constant flow of BHI was probed across both channels during these intervals. B, C and D represent three sodium azide (1 m%) injections for 4, 6 and 20 hours, respectively. Published in [87].

Table 2) was found at 10 Hz which is almost twice as much as obtained for 6-MHA.

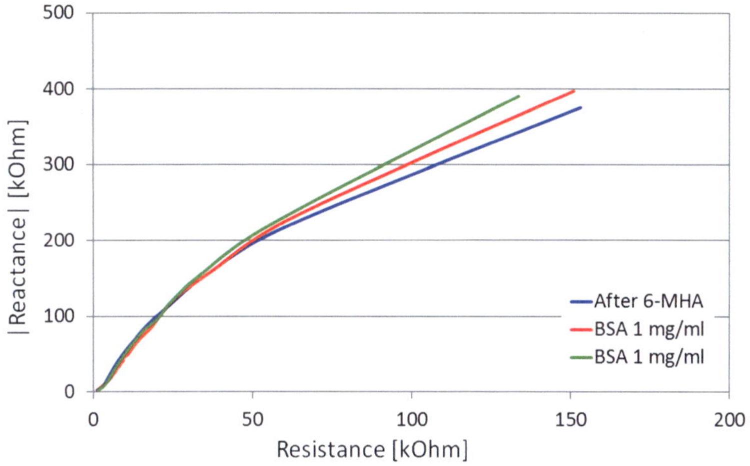

Figure 46 – Nyquist diagram of the nonspecific adsorption characterization of SAM layers created from 6-Mercaptohexanoic acid (6-MHA) by means of BSA adsorption experiments. All EIS measurements were preceded by a one minute electrolyte solution purge and carried out in electrolyte solution.

The last alkane SAM forming reagent tested was 11-MUA. This substance is very prominent in literature and used as modification layer for EIS. Examples include detection of cholera toxin [90] by coupling protein A, detection of infection markers by coupling protein G [91], and detection of tumor markers by directly immobilizing IgGs [92]. Its long alkane chain contributes to good self-alignment capabilities and high layer density due to stronger Van der Waals forces can be obtained. On the other hand, its lengthy chain is strongly electrically insulating. In the course of this work, freshly prepared SAM layers from 11-MUA showed initial impedance magnitude values of 1,6 MΩ at 10 Hz. This impedance is so high that EIS based detection could not be carried out. The reason why

11-MUA can be used by other researchers for creating layers of suffi-sufficiently low impedance is the fact that most of these systems use significantly lower frequency ranges (typically in the range of 50 mHz to 1 Hz). Chebil et al. [93] used frequencies down to 50 mHz, Chiriacò et al. [94] and Montrose et al. [91] used a frequency span down to 100 mHz during EIS based biosensing, Chen et al. [95] used frequencies of 1 Hz. In the course of this work, EIS was operated at higher frequencies in order to allow fast biosensing. Operating the measurement electronics at significantly lower frequencies would be an option to work with these high impedance layers, however, the measurement intervals would then be extended to several hours. This was deemed unsuitable and therefore lower impedance layers were sought.

One option tested was usage of an aromatic SAM-forming substance. 4-MBA, being the smallest aromatic thiol was deemed a suitable choice. Using this substance lower initial impedances layers should be achievable due to the presence of delocalized pi-electron systems in the benzene ring. This was verified experimentally (see Figure 47). Compared to 3-MPA (which was the SAM with the lowest impedance found) 4-MBA showed higher initial impedance values but slightly lower nonspecific adsorption. The still considerable nonspecific adsorption level for this compound can be attributed to its bulky characteristic, thus not allowing the formation of very densely packed layers.

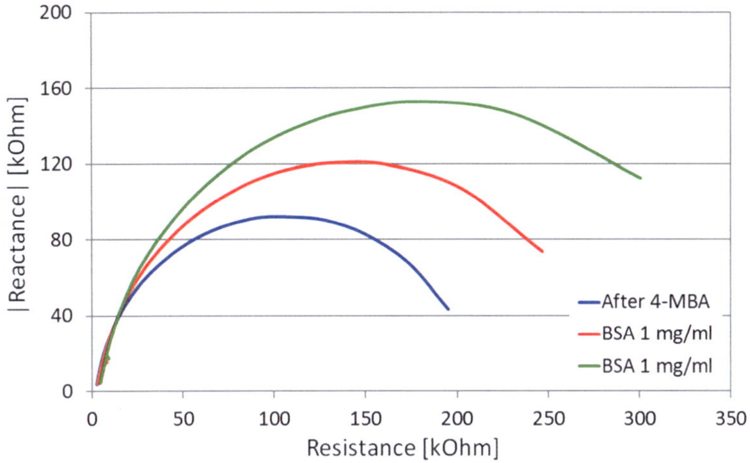

Figure 47 – Nyquist diagram of the nonspecific adsorption characterization of SAM layers created from 4-Mercaptobenzoic acid (4-MBA) by means of BSA adsorption experiments. All EIS measurements were preceded by a one minute electrolyte solution purge and carried out in electrolyte solution.

As an attempt to obtain a low impedance layer which also blocks the nonspecific adsorption, a combination of two different SAM-forming compounds was carried out. Initially, a solution containing equimolar amounts (2 mM) of 4-MBA and 3-MPA in ethanol was applied to the electrode surface with the objective to fill the spaces between the bulky 4-MBA molecules with 3-MPA. This approach yielded a low impedance layer but no significant improvement regarding the nonspecific adsorption. Mixing 4-MBA with 6-MHA resulted in a layer with significantly higher initial impedance (760 kΩ at 10 Hz). Therefore this combined SAM approach did not yield an improvement over usage of just a single SAM forming reagent.

In order to demonstrate the specific binding sensitivity of this method (see Figure 48) an electrode was modified with the best SAM-forming compound determined in the previous experiments (i.e., 4-MBA). After modification the carboxyl groups were activated by active ester chemistry (for details see Section 2.5.6). Then a streptavidin solution (125 µg/ml in PBS) was pumped at 50 µl/min into the channel and incubated on electrode surface for one hour, thus covalently immobilizing streptavidin to activated carboxyl group of 4-MBA via peptide bond formation (blue curve in Figure 48). PBS solution was pumped for 5 minutes at 50 µl/min in order to remove not bound streptavidin molecules from the surface. The electrode was then probed with a 300 µl BSA solution (25 µg/ml in PBS) at 50 µl/min to evaluate the degree of nonspecific adsorption (red curve in Figure 48). As can be seen, a slight shift in the curve can be observed indicating that nonspecific adsorption did occur. Subsequently biotin-modified BSA (biotinylated-BSA, b-BSA, 25 µg/ml, 300 µl at 50 µl/min) was probed across the electrode. Another PBS rinsing step was carried out (5 minutes at 50 µl/min) in order to remove loosely bound BSA molecules from the surface. Electrolyte solution was pumped into the system, allowing EIS measurements to take place. As can be seen (green curve in Figure 48) there is distinct signal increase. The value shift is bigger than the value obtained for nonspecific adsorption. However, as can also be seen, the signal shift is very small. The highest specific sensitivity achieved with this method was found to be 25 µg/ml of b-BSA using SAM layers from 4-MBA and streptavidin as covalently linked ligand. This limit is still orders of magnitude higher than the established threshold and therefore further surface modifications had to be investigated.

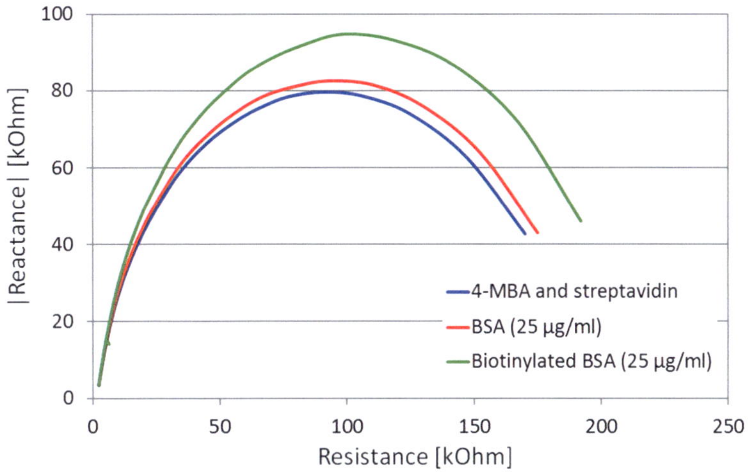

Figure 48 – Nyquist diagram of a specific binding assay of an electrode coated with 4-MBA. All EIS measurements were preceded by a one minute electrolyte solution purge and carried out in electrolyte solution.

4.4 Conductive polymer

The low sensitivity obtained by SAM-modified electrodes is mainly attributed to the insulating characteristic of these layers. This motivated investigating the usage of conductive polymers as interface layers. The polymer of choice in this work was polypyrrole due to its ease of synthesis (chemically or electrochemically), stability and high electrical conductivity. As discussed in Section 2.5.7 pyrrole derivatives with function groups suitable for ligand immobilization are commercially available. Therefore copolymers of pyrrole and pyrrole-3-carboxylic acid were investigated. Polypyrrole can be polymerized chemically (in solution) and electrochemically (on top of an electrode). Since the layers were required to be applied to the electrode anyway, polypyrrole was electrochemically polymerized directly on top of the measurement electrode. Electrochemical polymerization requires a three-electrode setup (see Section 2.1.4 for details) since it is very important to determine precisely the voltage being applied to the monomer solution.

Additional tests were carried out in order to assess the nonspecific adsorption levels on both pure polypyrrole and a copolymer containing a 3:1 molar ratio of pyrrole to pyrrole-3-carboxylic acid.

In a typical polymerization process, a solution containing 120 mM of pyrrole, 40 mM of pyrrole-3-carboxylic acid and 100 mM of potassium chloride (required to increase the solution conductivity) was prepared. This solution was probed into the PDMS low cell thus probing the electrodes. Polymerization was initiated by applying a triangular shape excitation signal ranging from -800 mV to 800 mV at a speed of 40 mV/s [96]. Preliminary experiments were carried out with the aim of determining the optimized number of cycles as a compromise between layer thickness (thicker layers result in higher initial impedance) and layer homogeneity (holes in the layer enable high nonspecific adsorption). The optimal number of cycles was determined to be three.

The formed polymer could be visually seen as a dark layer forming on top of the working electrode electrodes. Experiments done in order to optimize the polymer film homogeneity (characterized visually in the light microscope) under various flow rates (ranging from 50 µl/min to 0.5 µl/min) during the electropolymerization did not show to have considerable influence upon the quality of the polymerized layers.

In order to assess the layer characteristics fluorescence microscopy (see Section 2.9 for details) was used. A fluorescence marked (alexa fluor® 488 dye) antibody (rabbit anti-goat IgG) was used to evaluate the nonspecific adsorption levels. The results can be seen in Figure 49. The electrode coated with pure polypyrrole (Figure 49 a) shows higher nonspecific adsorption levels as polypyrrole provides a non-polar surface to which the antibodies can adhere by hydrophobic interaction. The electrode coated with the copolymer (Figure 49 b) shows lower nonspe-nonspecific adsorption levels. Further experiments with higher concentrations of pyrrole-3-carboxylic acid led to very inhomogeneous and high impedance polymer films. Experiments trying to polymerize pure pyrrole-3-carboxylic acid solutions failed as it was found to be impossible to polymerize this solution.

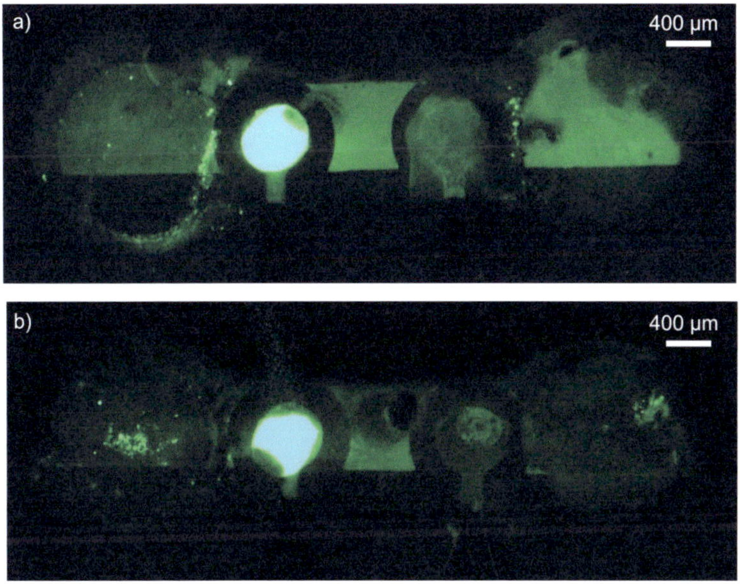

Figure 49 – Fluorescence microscopy images of two electrodes coated with two different polypyrrole (co)polymers. Each electrode was incubated for 60 minutes in a 100 µg/ml solution containing rabbit anti-goat IgG marked with alexa-488. a) Electrode coated with 100% pyrrole. b) Electrode coated with a copolymer with 1:3 molar ratio of pyrrole-3-carboxylic acid to pyrrole. The bright spot on both electrodes represents the reference electrode which consists of Ag/AgCl paste. This paste is very porous and adsorbs promptly the secondary antibody, creating the bright spots.

Two different substances (TRIS and an amine-functionalized PEG - α,ω-Bis-amino PEG with a molecular weight of 2000 g/mol) were coupled to the carboxyl groups on the copolymer's surface in order to reduce the nonspecific adsorption (see Figure 50). As discussed in [48] increasing the hydrophilicity of the surface should reduce the potential of hydrophobic interaction and therefore the nonspecific adsorption. As can be seen TRIS (see Figure 50 b) did result in significant reduction of nonspecific adsorption. Furthermore, the PEG-modified layer also shows significantly reduced nonspecific adsorption (see Figure 50 c). However, initial experiments with EIS using PEG-modified polypyrrole copolymers showed initial impedances of over 1 MΩ at 10 Hz which renders these surfaces unsuitable for analytical applications.

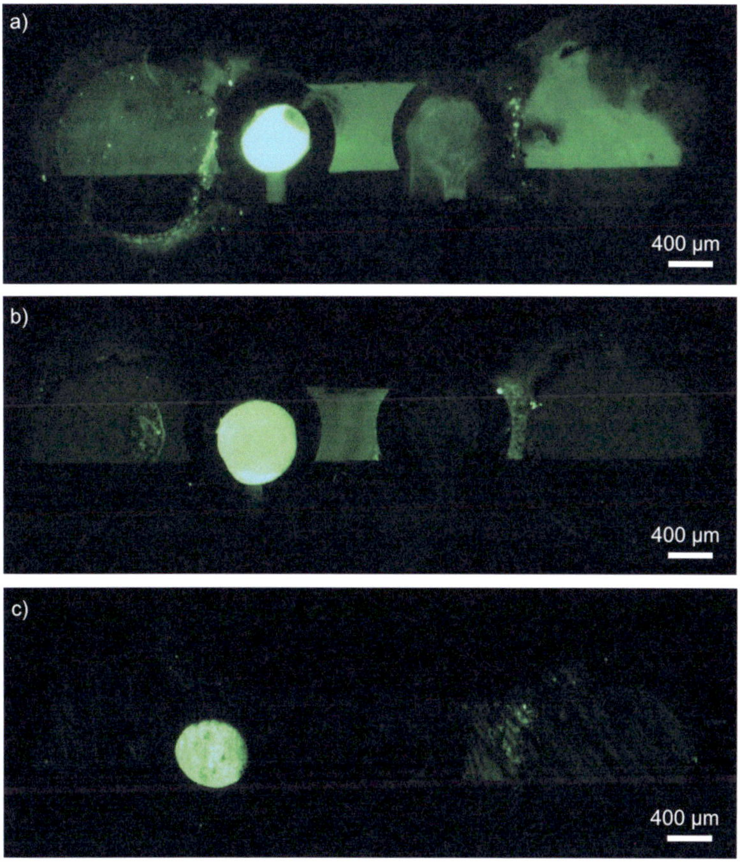

Figure 50 – Comparison of fluorescence microscopy pictures of two different nonspecific adsorption blocking compounds coupled a copolymer with 1:3 molar ratio of pyrrole-3-carboxylic acid to pyrrole. After the coupling of these compounds the electrode was incubated for 60 minutes in a solution containing 100 µg/ml of antibody (rabbit anti-goat IgG marked with alexa-488). a) Electrode coated with pure polypyrrole from Figure 49 a as control. b) Coupling of TRIS to the copolymer surface. c) Coupling of an amine-functionalized PEG (α,ω-Bis-amino PEG with a molecular weight of 2000 g/mol).

In the next step, EIS experiments were carried out using BSA as a block-blocking agent in order to assess the nonspecific adsorption. Three cycles polymerizing a 3:1 molar ratio solution of pyrrole and pyrrole-3-carboxylic acid (see Figure 51 a) resulted in the best compromise between layer thickness and layer homogeneity established by EIS-based characterization.

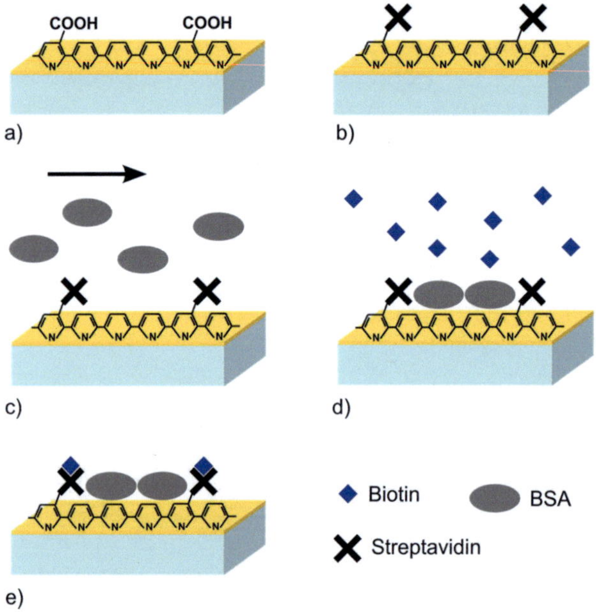

Figure 51 – Schematic diagram of the surface modification based on polypyrrole and the assay performed. a) After the electropolymerization of the copolymer (3:1 pyrrole to pyrrole-3-carboxylic). b) Immobilization of streptavidin by means of active ester chemistry. c) BSA blocking of the surface in order to reduce nonspecific adsorption (d). e) Specific detection of biotin.

For this, substrates were prepared and coated by electropolymerization as described (see Figure 51 a). After the electropolymerization streptavidin was immobilized via active ester chemistry (EDC/NHS) (see

Section 2.5.6 for details) onto the polymer surface (see Figure 51 b). Subsequently solutions of 1 mg/ml BSA in PBS (300 µl volume at 50 µl/min) were probed through the flow cell with the aim of assessing the unspecific adsorption and blocking subsequent samples from interacting nonspecifically with the electrode (see Figure 51 c). Samples containing biotin (either conjugated to BSA or in pure form) were used to test the sensor's sensitivity to the specific binding of a high and low molecular weight analyte, respectively (see Figure 51 d and e). The high molecular weight conjugated analyte b-BSA (biotinylated BSA, 66.500 g/mol) could be detected at a detection limit of 1 ng/ml which is in the expected range for analytical applications. Based on this result an analogous assay was carried out with the aim of specifically detecting biotin (see Figure 52) which is a low molecular weight analyte (244 g/mol). The impedance was measured after the electropolymerization (see blue curve in Figure 52), after immobilization of streptavidin via active ester chemistry (see red curve in Figure 52), after two blocking injections of 1 mg/ml of BSA in PBS (see green and yellow curve in Figure 52), and after probing of a sample containing 10 and 100 ng/ml of biotin in PSB (see purple and black curve in Figure 52).

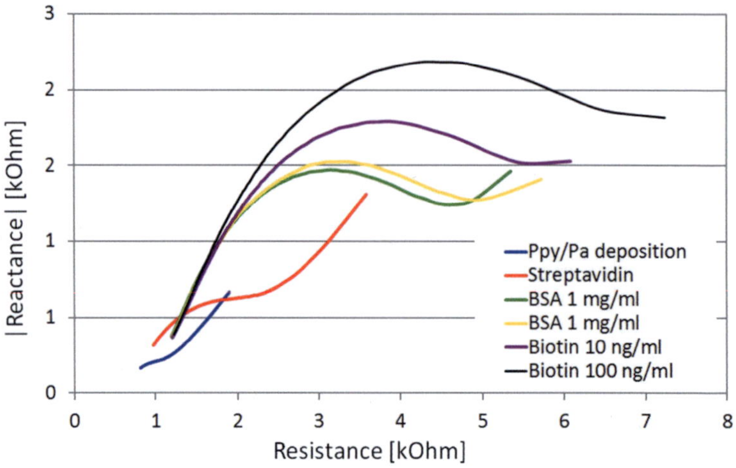

Figure 52 – Nyquist diagram of a specific binding assay of an electrode covered with a conductive polypyrrole copolymer (pyrrole and pyrrole-3-carboxylic acid, 3:1 molar ratio). All EIS measurements were preceded by a one minute electrolyte solution purge and carried out in electrolyte solution.

Surface modifications based on the described approach using copolymers of polypyrrole allowed significant increases of sensor sensitivity. For the best SAM a limit of detection of 25 µg/ml of b-BSA was achieved. Using polypyrrole copolymers this value was reduced by 4 orders of magnitude down to 1 ng/ml. However, the nonspecific adsorption is still considerably high as can be seen by the increase in impedance after BSA probing in Figure 52. This is mainly due to the fact that polypyrrole, being a nonpolar polymer, gives rise to significant hydrophobic interactions. However, experiments using reagents for reducing nonspecific adsorption (TRIS and amine-functionalized PEG - α,ω-Bis-amino PEG with a molecular weight of 2000 g/mol) coupled via active ester chemistry did not result in significant improvements. This may be due to the fact that the molar ratio of pyrrole to pyrrole-3-

carboxylic acid was found to be at an optimum at 3:1 molar ratio which results in polymers with a high percentage of nonpolar pyrrole. Increasing the amount of functionalized monomers led to a gradual decrease of the layer conductivity. Creating polymers solely from pyrrole-3-carboxylic acid failed due to the fact that the monomer did not form by electropolymerization.

Since increasing the concentration of surface functional groups beyond a 1:3 molar ratio was impossible using polypyrrole copolymers another surface modification method was investigated which allowed creating functional layers with stoichiometric amounts of function groups.

4.5 Chemical grafting

In order to create a surface modification which offers a high density of functional groups chemical grafting was chosen. This method enables the direct immobilization of small molecules to an electron rich surface (for details see Section 2.5.4) by means of cyclic voltammetry.

Initially an aqueous solution (8.3 ml) containing 2 M of sodium nitrite and a 0.1 M 4-aminobenzoic acid (4-ABA) in 0.5 M HCl solution (50 ml) were prepared. 4-ABA consists of a benzene ring substituted with an amino and carboxyl group at opposite positions, thus providing the required groups (amino group for the chemical grafting and carboxylic group for posterior functionalization) and a delocalized pi-electron system which increases the grafted layer's conductivity. Both solutions were initially stirred and cooled down to 5 °C in an ice bath. Slowly adding the sodium nitrite solution to the 4-ABA solution resulted in slight temperature increases and a shift of solution color to light yellow. This procedure was repeated until both solutions were completely mixed. The temperature of this solution was tightly controlled at 5 °C throughout this procedure in order to avoid the rapid emission of nitrogen gas and to avoid the deterioration of the highly instable aryl diazonium compounds (see Section 2.5.5 for details). The solution was stirred for 20 minutes at 5 °C after which the whole electrode was submerged in the solution and cyclic voltammetry was performed for two cycles from 0,2 V to -0,6 V at a speed of 50 mV/s. A clear color change was observed on the working electrode as it acquired a darker tint. The number of cycles was a compromise between electrode corrosion (resulting from electrochemical oxidation of the working electrode in acidic media) and electrode color change (as an indicator of the successful chemical grafting procedure) evaluated optically. During this procedure the creation of gas (nitrogen) bubbles was observed as sign that the reaction has taken place. After the cycles were completed the electrode was thoroughly washed in bidistilled water and dried under nitrogen flow. The electrode was then inserted into the flow cell for further surface functionalization.

The preparations for this method, e.g., weighing all substances, cooling and stirring, are time consuming. The tight temperature control requires constant monitoring. Trying to reduce the time required for cooling down the solution and the amount of necessary reagents by reducing the total solution volume and performing the reaction inside the flow cell resulted in two disadvantages. Firstly, the flow cell hinders the dissipation of heat resulting from the chemical grafting, thus destabilizing the reaction. Secondly, the nitrogen gas produced during the reaction fills up the microchannel thus interrupting the reaction. Early experiments carried out in the flow cell showed low electrode-to-electrode reproducibility. Pumping the grafting solution during the cyclic voltammetry did not improve the reproducibility significantly.

Again, fluorescence microscopy (see Section 2.9 for details) was used in order to assess the characteristics of the layers created. Using fluorescence microscopy two cases were characterized in order to verify that the created layer fulfilled both requirements of an ideal surface modification, i.e., binding of the analyte of interest (positive control, see Figure 53) while suppressing the nonspecific interaction (negative control, see Figure 54).

The assay used to verify the capability of the layer to immobilize the ligand of choice was based on the immobilization of a primary antibody which binds to a fluorescently labeled secondary antibody. This assay configuration was chosen as it demonstrates that the immobilization procedure based on active ester chemistry can be carried out as expected. In these experiments the ligand immobilized by active ester chemistry was a rabbit anti-goat IgG which was used as the primary antibody (see Figure 53). The secondary antibody was a fluorescently labeled goat anti-rabbit IgG. A solution containing 100 µg/ml of the primary antibody in PBS was incubated on the freshly grafted electrodes for 60 min. Incubation was performed using the flow cell and the microfluidic channel in order to reduce sample evaporation. The fluorescence label present in the secondary antibody was Alexa fluor® 488 dye (absorption and maximum maxima at 495 nm and 519 nm, respectively). After coupling the primary antibody the electrode was thoroughly washed with PBS and incubated in a solution containing 1 µg/ml of secondary antibody in 5 m% BSA in TBST (TRIS-Buffered Saline and Tween 20 – 50 mM Tris, 150 mM NaCl and 0.05 vol% Tween 20) overnight. Afterwards the electrode was thoroughly washed in PBS before being examined under the fluorescence microscope.

The results of this assessment are shown in see Figure 55. As expected, the electrode shows up as a bright rectangular area in the form of the microfluidic channel used during probing (see Figure 55 a). This demonstrates that the layer is capable of effectively binding the analyte in question.

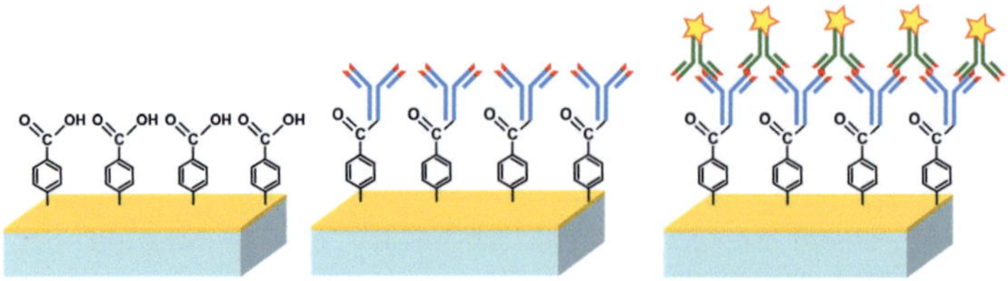

Figure 53 – Schematic diagram of an antibody immobilization process onto a grafted layer of 4-Aminobenzoic acid (4-ABA) and subsequent coupling of the primary antibody by means active ester chemistry. The secondary antibody is fluorescently labeled with alexa-488 and binds to the primary antibody (incubated overnight in TBST and 5 m% BSA). This surface preparation protocol generates electrodes which show strong fluorescence if binding occurs (positive control).

The second assay used to test the suitability of the created layers was implemented with the aim of assessing the degree of nonspecific adsorption. This assay is based on the immobilization of TRIS (see Section 2.5.3 for details) by means of active ester chemistry. For this an aqueous solution (300 µl) of TRIS 100 µg/ml was incubated in the flow cell for 60 minutes. After probing the surface is thoroughly rinsed with bidistilled water. Creating a dense TRIS-modified surface layer renders the electrode hydrophilic. As discussed hydrophilic surfaces are less prone to nonspecified adsorption by means of hydrophobic interaction (see Figure 54). Afterwards, a solution containing 1 µg/ml in TBST and 5 m% of the secondary antibody was incubated in the flow cell overnight.

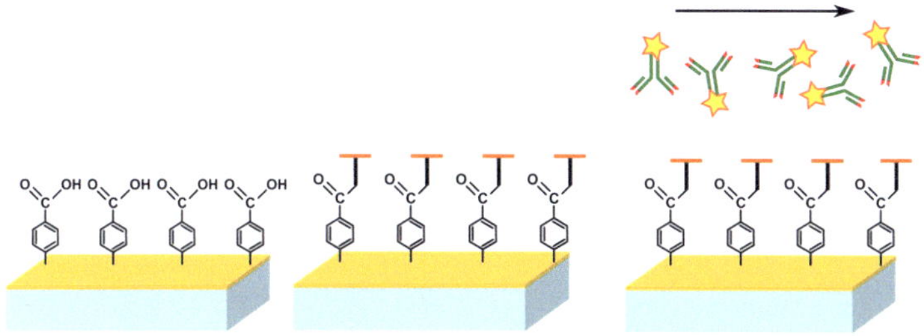

Figure 54 – Schematic diagram of the suppression of nonspecific adsorption using TRIS coupled to a grafted 4-Aminobenzoic acid layer (4-ABA). After the grafting of 4-ABA the carboxyl groups were activate by active ester chemistry and TRIS was coupled onto the surface. The secondary antibody is fluorescently labeled with alexa-488 and has no specific affinity to TRIS. This surface preparation protocol generates electrodes which show low fluorescence due to the absence of adsorbed secondary antibody (negative control).

The result of this assay can be seen in Figure 55 b. The TRIS-modified electrode remained considerably darker than the electrode modified with the primary antibody. This indicates that the hydrophilic TRIS layer significantly reduces adsorption effects thus decreasing the fluorescence signal. In order to further decrease the nonspecific binding a SAM forming reagent, i.e., 11-Mercaptoundecanol (11-MUE) a polar alkane thiol was used (2 mM solution in ethanol overnight). Adding this SAM forming reagent allowed blocking areas of the electrode with inhomogeneously grafted surface layers. Using this mixed grafted/SAM layer it was possible to further improve the suppression of the nonspecific adsorption as can be seen in Figure **55** c.

In all images, the porous Ag/AgCl paste (required for the cyclic voltammetry during grafting) can be seen brightly (see white circle). The paste absorbs the secondary antibody strongly thus creating the bright spots seen in the images.

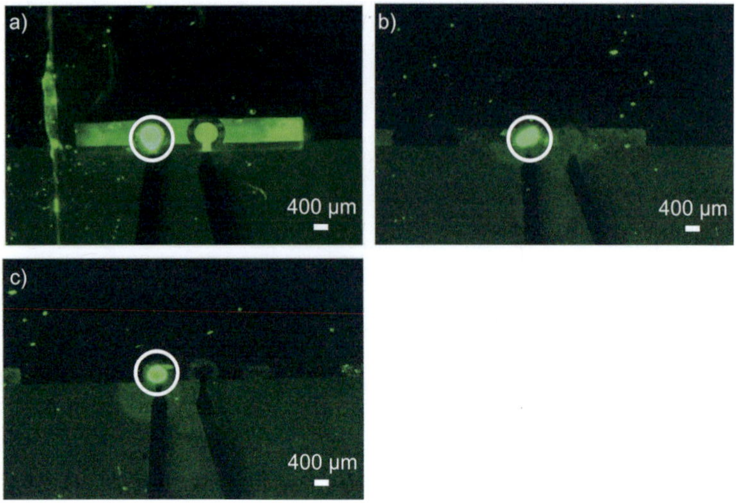

Figure 55 – Fluorescence microscopy pictures of three different surface preparations after the grafting of 4-Aminobenzoic acid layer. a) The immobilization of the primary antibody (100 µg/ml in PBS for 60 minutes) – positive control. b) The coupling of TRIS (100 µg/ml aqueous solution for 60 minutes) to the grafted (negative control). c) Combined grafting/MUA layer extending the modification show in b. White circles mark the Ag/AgCl paste used for the reference electrode.

Unfortunately, a bad electrode-to-electrode reproducibility was obtained among the fluorescence microscopy experiments due to the instability and sensitivity of the electrochemical grafting process. Even tightly controlling the temperature and stirring did not lead to a considerable improvement.

EIS measurements confirm the results obtained by fluorescence microscopy. The partial suppression of the nonspecific adsorption on surfaces created from grafting only can be seen in Figure 56. The blue curve shows the bare electrode impedance (see Figure 56). The considerable increase seen in the red curve (see Figure 56) is due to the high concentration of carboxylic acids which protonate leaving the surface highly negatively charged. This in terms reduces the current flow thus increasing the impedance. After TRIS coupling (100 µg/ml in bidistilled water for 60 minutes) the carboxylic groups are bound to the ligand thus the surface charge reduces and, consequently, the impedance reduces as well. TRIS is coupled by means of active ester chemistry using EDC/NHS. The impedance measured after TRIS coupling can be seen in the green curve (see Figure 56). The impedance measured after incubating the electrode in a solution with 100 µg/ml in PBS of the primary antibody is shown in the yellow curve. This experiment confirms the results obtained from fluorescence microscopy seen in Figure 55 b. Even coupling TRIS to the grafted electrode does not completely suppress the nonspecific adsorption which is represented by the impedance increase between the green and yellow curves of Figure 56.

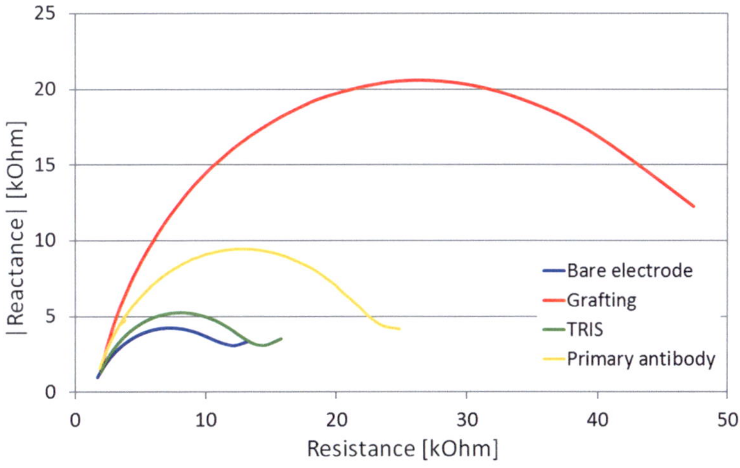

Figure 56 – Nyquist diagram of the nonspecific adsorption characterization of an electrode modified by grafting of 4-Aminobenzoic acid (4-ABA).

EIS experiments combining chemical grafting with SAM showed no significant improvement of the high nonspecific adsorption seen in grafted layers.

Fluorescence microscopy experiments showed no considerable improvement of layer homogeneity with tight temperature control and stirring, thus leading to high nonspecific adsorption. Initial attempts to perform the reaction in the flow cell (in stationary and in flow systems) and therefore to reduce the total volume of the grafting solution and accelerate the solution cooling down led to no improvement.

4.6 Photobleaching

The final surface modification method uses a novel approach based on photobleaching fluorophore-containing ligands to an unsaturated surface, in this case gold coated with allyl mercaptan. Holden and Cremer [56] showed how to use photobleaching to immobilize ligands coupled with a fluorophore to BSA-coated surfaces. Waldbaur et al. [57] used this technique to create protein patterns using visible light structured using a maskless projection lithography system (see Section 2.5.5 for details). This approach does not require the use of coupling methods such as active ester chemistry to immobilize the ligands to the surface. Ligands labelled with fluorescein can be acquired pre-purified to ensure the label does not interfere with the ligand's binding sites. To the best of the author's knowledge this work is the first to report usage of photobleaching as ligand immobilization technique.

The first step is to create the unsaturated surface by means of SAM formation. Initial experiments using allyl mercaptan solubilized in ethanol and methanol did not produce monolayers compatible with this method. Due to the high vapor pressure of allyl mercaptan it is possible to create the SAM in the gas phase. For this about 1 ml of allyl mercaptan was pipetted into a glass vial which was placed in a desiccator together with the electrode substrate. The desiccator was sealed and kept in the dark overnight allowing formation of the SAM (see Figure 57 a). After washing the substrate with acetone and ethanol and drying with nitrogen the photobleaching procedure was carried out.

In order to determine the effectiveness of this method two different surfaces had to be created by means of photobleaching as positive and negative control. The successful immobilization of a ligand to the surface (see Figure 57 c, positive control) and the creation of a surface which suppresses nonspecific adsorption (see Figure 57 e, negative control) can be created using this method. These surfaces were created by photobleaching fluorescein-biotin and fluorescein-TRIS, respectively. Fluorescein-TRIS stock solution was obtained by mixing equimolar amounts (2 mM) of FITC and TRIS in 1 ml dimethyl sulfoxide (DMSO) and vigorous agitation overnight. All further dilutions were done by adding PBS to the stock solution. The isothiocyanate group is reactive toward the nucleophilic primary amine present in TRIS resulting in the formation of a fluorescein-TRIS compound. As described in Section 2.5.3 it is very important for EIS to generate surfaces sparsely immobilized with ligands. The dense immobilization of ligands could lead to high impedance values and therefore to low sensor sensitivities. A fluorescence scanner (Gene Pix 4000B from Axon Instruments) was used to complement the analysis of the electrodes modified using this method.

In a typical photobleaching modification, 30 µl of an 80 µM fluorophore solution in PBS was pipetted onto the electrode and covered with a glass coverslip in order to reduce evaporation. The surface was irradiated using the UV system used for lithography. Fluorescein (excitation maximum at 494 nm) was bleached using filter position 7 (450 nm to 535 nm with maximum at 487 nm, see Section 3.1.4). The bleaching was carried out at maximum intensity (approximately 4 mW/cm^2) for 10 minutes. Afterwards, the substrate was washed with ethanol and bidistilled water and agitated in a chamber with 6 ml of PBS three times for 5 minutes each. The substrate was then washed with bidistilled water and dried with nitrogen.

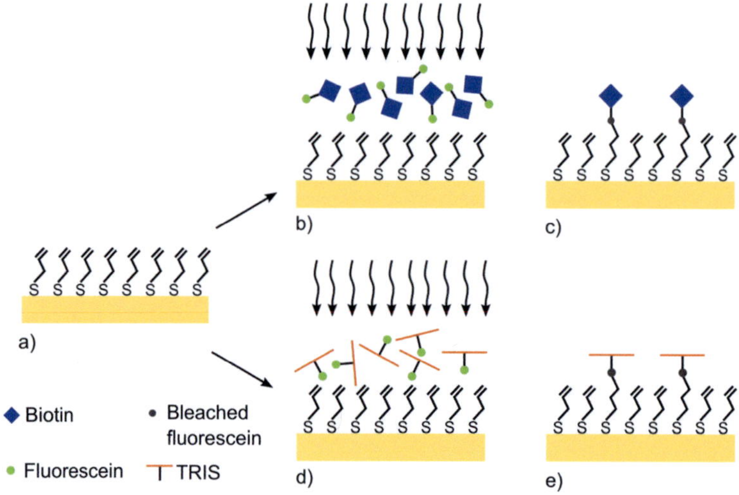

Figure 57 – Schematic diagram of the photobleaching-based immobilization technique. a) A SAM of ally mercaptan formed in the gas phase. b) Fluorescein-biotin is photobleached. c) Immobilization of biotin to the non-saturated surface (positive control). d) Fluorescein-TRIS is immobilized by photobleaching. e) Immobilization of TRIS to the unsaturated surface (negative control).

As fluorescein is bleached during the immobilization and thus cannot be detected directly in the fluorescence scanner the electrode was stained in the flow cell using 30 µl of a solution containing 5 µg/ml of streptavidin coupled to cyanine-3 (excitation from 450 to 580 nm with maximum at 552 nm) per channel. The solution was incubated for 60 minutes in the dark (see Figure 58 a). Streptavidin-cy3 was also used for assessment of nonspecific absorption on the TRIS-coated electrodes (negative control). If the layers are sufficiently non-adsorbing streptavidin should not adsorb on the hydrophilic surface and thus the electrode should remain dark in the fluorescence scanner images (see Figure 58 b). After staining the flow cell was removed and the electrode was washed with bidistilled water and agitated in a chamber with PBS for 5 minutes three times. The substrate was then placed in the fluores-fluorescence scanner and scanned.

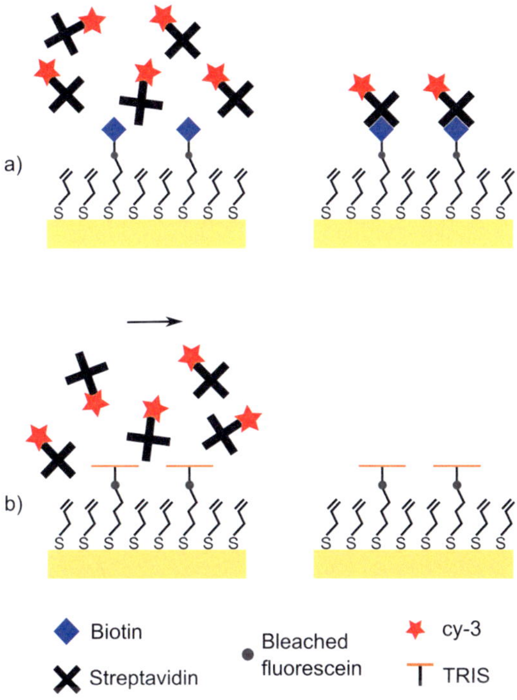

Figure 58 – Schematic diagram of the staining protocols using streptavidin-cy3. a) The biotin coated surface allows the specific binding to streptavidin-cy3, thus creating brightly fluorescent areas. b) The TRIS coated surface blocks nonspecific adsorption of streptavidin-cy3 thus remaining dark during fluorescence microscopy.

Fluorescence scanner images confirmed the expectations for these surface preparations (see Figure 59). The images were taken using the fluorescence scanner (532 nm, photon multiplier voltage 500 V, power 100%, pixel size 10 µm, 1 line per average). In Figure 59 a the bright area in the form of the microfluidic channel used during staining can be clearly seen confirming that streptavidin-cy3 was bound to the biotin modified surface. The strong fluorescence exhibited by the glass surface can be attributed to the nonspecific adsorption of proteins, in this case streptavidin-cy3, due to electrostatic attraction [97]. Figure 59 b on the other hand illustrates that the TRIS modified surface was able to block the nonspecific adsorption of streptavidin-cy3. Very low fluorescence is seen on the glass surface due to the nonspecific adsorption of fluorescein-TRIS. Figure 59 c demonstrates the control electrode which was not modified and consists solely of an allyl mercaptan SAM. One can clearly see the low nonspecific adsorption of streptavidin-cy3 to allyl mercaptan functionalized gold surface in contrast to the high nonspecific adsorption of streptavidin-cy3 onto glass. This experiment reinforces the successful coupling of fluorescein-biotin to the allyl mercaptan surface from Figure 59 a as the brightness of the gold surface cannot be attributed to nonspecific adsorption of streptavidin-cy3.

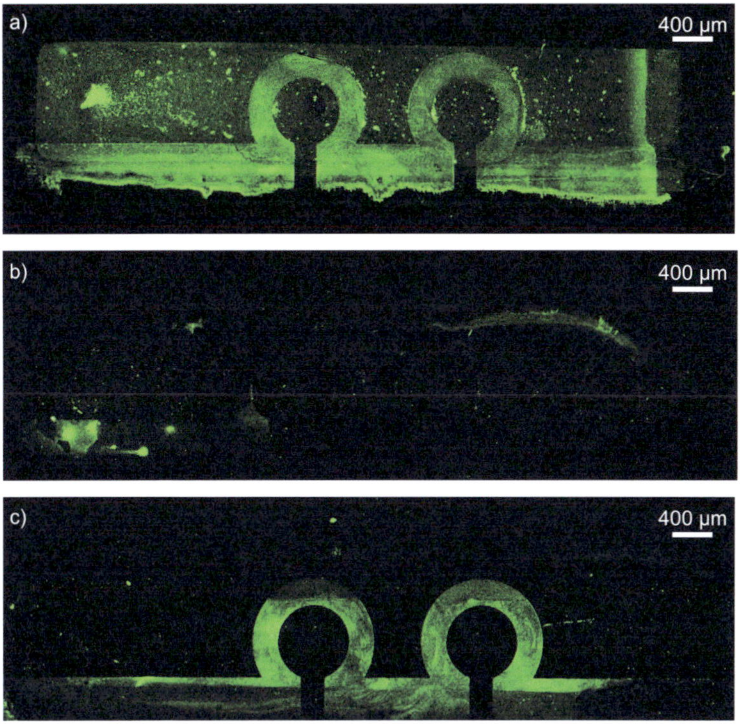

Figure 59 – Fluorescence scanner images of three surfaces prepared during the photobleaching protocol. The staining solution was 30 µl of a 5 µg/ml solution of streptavidin-cy3 in PBS per channel. a) Image of an allyl mercaptan SAM to which biotin was coupled using photobleaching. b) Image of an allyl mercaptan SAM to which TRIS was coupled using photobleaching. c) Control electrode which was not modified after the formation of allyl mercaptan SAM.

After the initial layer characterization tests using fluorescence staining, experiments using impedance spectroscopy were performed. Since the coupling of TRIS to the surface blocked the nonspecific adsorption considerably it was incorporated into the electrode functionalization protocol.

Initially 30 µl of an 80 µM of the fluorescein-biotin solution in PBS was pipetted onto the electrode and covered with a glass coverslip in order to reduce evaporation. The surface was bleached using filter position 7 at maximum intensity (approximately 4 mW/cm^2, see Section 3.1.4) for 10 minutes. Afterwards, the substrate was washed with ethanol and bidistilled water and agitated in a chamber with 6 ml of PBS three times for 5 minutes each. The substrate was then washed with bidistilled water and dried with nitrogen (see Figure 60 a). The same exposure and washing procedures as described for fluorescein-biotin were also used for coupling fluorescein-TRIS (see Figure 60 b) in order to further hinder nonspecific adsorption. After this second bleaching, the electrode is inserted into the flow cell and the first EIS measurement is carried out. Afterwards, each electrode was incubated in 30 µl of a solution containing 5 µg/ml of streptavidin-cy3 in the flow cell for one hour followed by 1 minute rinsing with PBS and recording of an EIS spectrum (see Figure 60 c and d).

In order to assess (and subsequently block) the nonspecific adsorption all electrodes were probed at 50 µl/min with a 300 µl of a 1 mg/ml BSA in PBS solution in the flow cell before being washed with PBS for 1 minute in the flow cell. An EIS spectrum was subsequently recorded. This procedure was repeated twice. Subsequently the samples containing different concentrations of biotin in PBS (see Figure 61 b and c) or b-BSA in PBS (see Figure 61 d and e) were probed across the electrode surface (300 µl at 50 µl/min). After each probing an EIS spectrum was recorded.

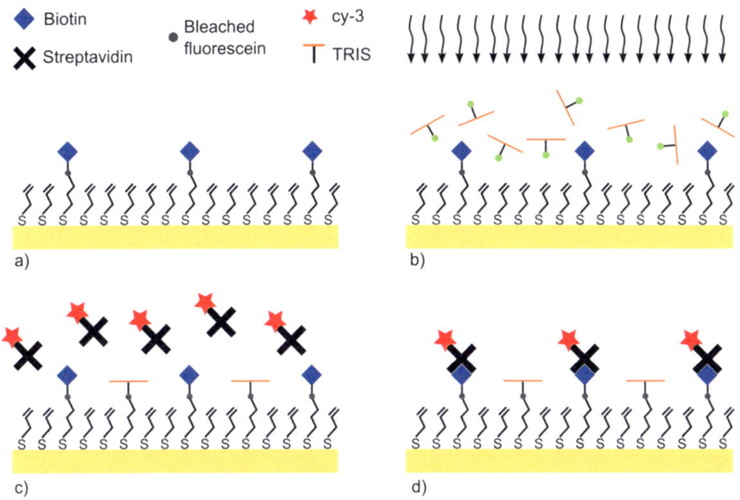

Figure 60 – Schematic diagram of the surface preparation protocol using photobleaching. a) Fluorescein-biotin is first coupled by means of photobleaching on the allyl mercaptan functionalized surface. b) Subsequently fluorescein-TRIS is immobilized. c) The electrode is placed into the flow cell and streptavidin-cy3 is probed across the electrode resulting in specific binding to the biotinylated surface (d).

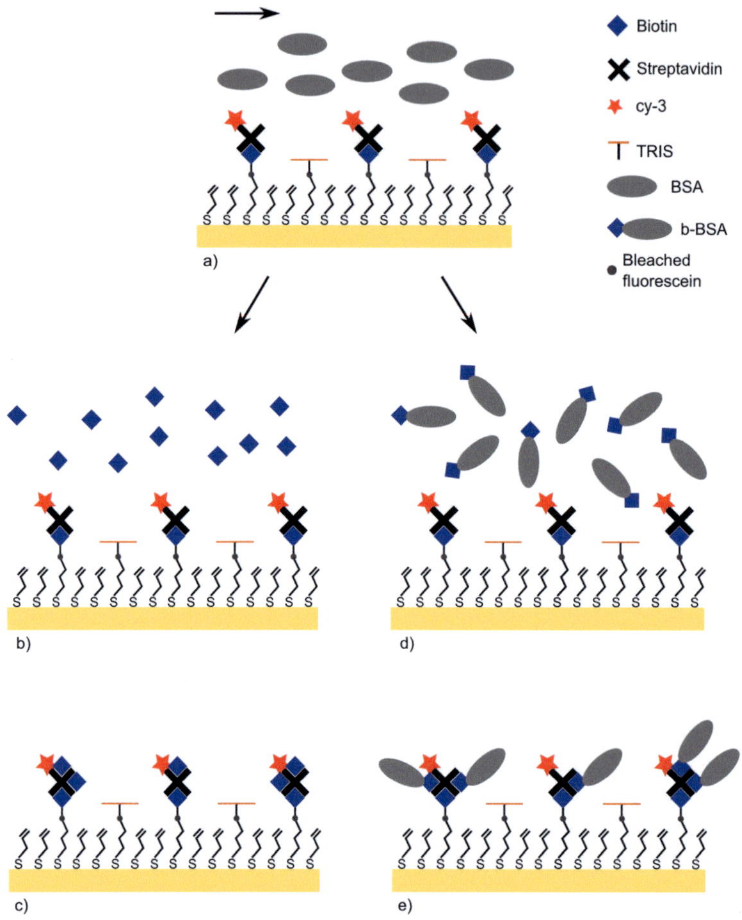

Figure 61 – Schematic diagram of two EIS assays (continuation of Figure 60) for the detection of biotin and b-BSA. a) Electrodes are probed with 300 µl of a 1 m/ml BSA in PBS solution in order to assess nonspecific adsorption. b) The electrode is then probed with solutions containing various concentrations of biotin in PBS (300 µl at 50 µl/min). c) Biotin binds specifically to the surface-bound streptavidin/biotin complex. d) The electrode is probed with a solution containing various concentration of b-BSA in PBS. e) b-BSA binds specifically to the surface-bound streptavidin/biotin complex.

Figure 62 (blue curve) exemplifies that the impedance of the electrode modified by photobleaching still remains relatively small. The addition of streptavidin-cy3 (see red curve in Figure 62) results in a considerable increase as expected due to this molecule's size. The first probing with BSA (see green curve in Figure 62) shows that the layer still shows nonspecific adsorption but the second BSA probing (see yellow curve in Figure 62) confirms that the adsorption quickly saturates as this second probing causes almost no further signal increase. The first experiment (Figure 62) was carried out using b-BSA as analyte. After probing the electrode with 1 ng/ml of b-BSA in PBS (see purple curve in Figure 62) a clear signal increase can be observed. Further sample injections at concentrations of 10 ng/ml and 100 ng/ml of b-BSA (see black and orange curves in Figure 62) could also be successfully detected.

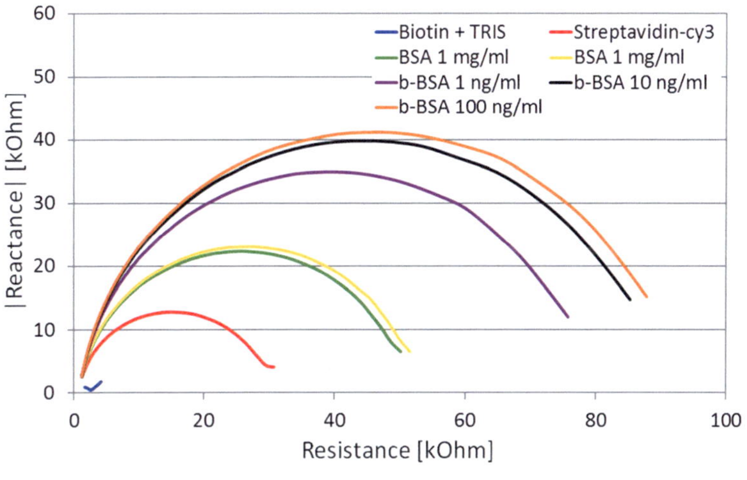

Figure 62 – Nyquist diagram of electrode modified by means of photobleaching. These measurements were performed in order to assess the limit of detection for b-BSA. All EIS measurements were preceded by a one minute electrolyte solution purge and carried out in electrolyte solution.

A second analogous experiment was carried out in which b-BSA was replaced by a low molecular weight analyte, biotin (see Figure 63). The preparation steps prior to the analyte probing were identical. As can be seen, there is a high correlation between every curve from a) and b), starting from the initial measurement up to the final biotin 10 ng/ml probing. Not only are the curve shapes almost identical but the impedance increase between the steps are consistent. This indicates a good electrode-to-electrode reproducibility. It is also important to highlight the impedance increase between the purple and black curves from Figure 63, it can be seen that even though the concentration of the analyte has been increase 10 fold the sensor's response is quite small, thus indicating a saturation of binding sites on the electrode's surface. This can probably be attributed to the sparsely functionalized (with analyte specific molecules, e.g., biotin) surface which was required to maintain the sensor's initial impedance low.

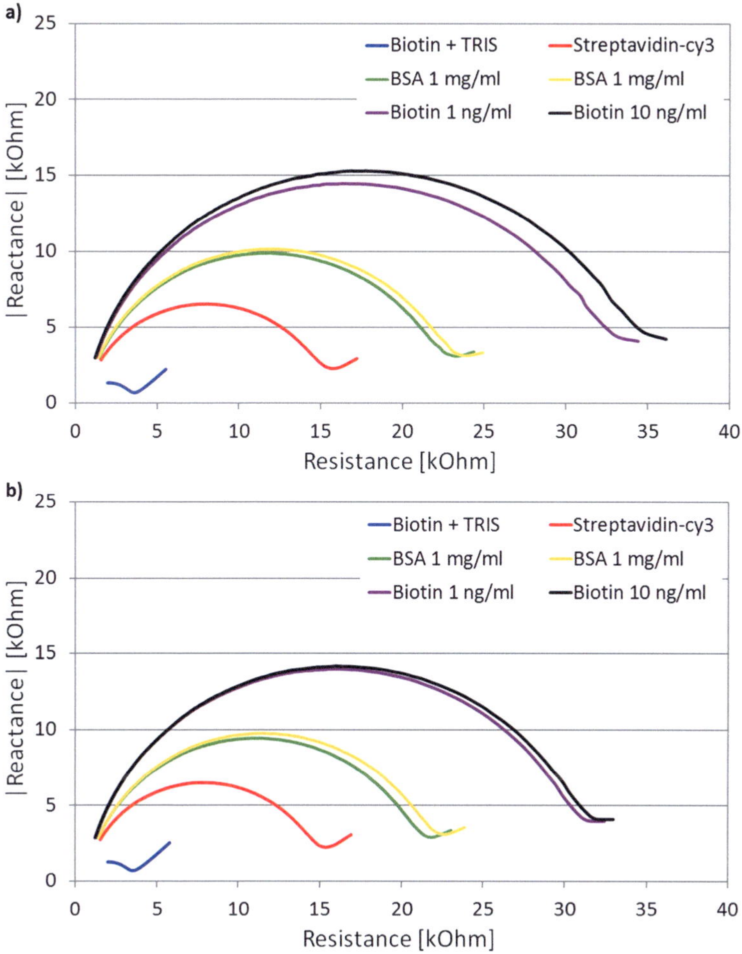

Figure 63 – Nyquist diagrams of two different electrodes modified by means of photobleaching. These measurements were performed in order to assess the limit of detection for biotin. All EIS measurements were preceded by a one minute electrolyte solution purge and carried out in electrolyte solution. As can be seen, there is a high correlation between the curves from a) and b), indicating a good electrode-to-electrode reproducibility.

A final experiment based on this surface modification by means of photobleaching was carried out in order to determine the detection limit for biotin. For this, surface modification was carried out as described (see Figure 60). After the immobilization of streptavidin-cy3 to the electrode's surface three subsequent injections of 300 µl of a 1 mg/ml BSA in PBS at 50 µl/min were probed across the electrode (see green, yellow and purple curves in Figure 64) in order to assess and block nonspecific adsorption. As can be seen from Figure 64 the blocking is complete after the second injection as the two curves cannot be distinguished (the purple curve lies on top of the yellow curve). Subsequently, a control injection of pure PBS (zero sample, 0 ng/ml, see black curve in Figure 64) was probed across the electrode. The next injection was a sample containing 10 pg/ml of biotin in PBS which resulted in a small impedance increase which is very similar to the shift of the zero sample and thus cannot be reliably detected (see orange curve in Figure 64). The next sample injected was a 100 pg/ml biotin in PBS solution and generated a much higher signal increase (see light blue curve in Figure 64). The final sample injected contained a concentration of 1 ng/ml of biotin in PBS and can, again, be clearly distinguished (see light red curve in Figure 64).

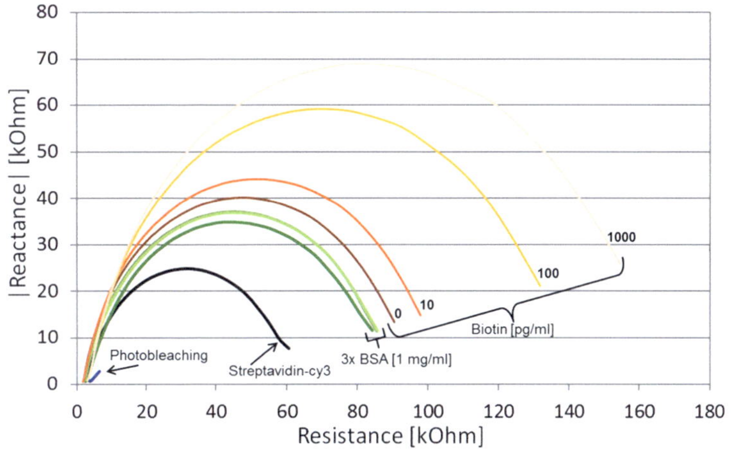

Figure 64 – Nyquist diagram of an electrode modified by means of photobleaching. This assay was performed in order to further investigate the limit of detection for biotin. All EIS measurements were preceded by a one minute electrolyte solution purge and carried out in electrolyte solution.

Surface functionalization based on photobleaching of fluorophore-containing ligands to an unsaturated surface is a new and powerful technique for electrode surface modification. The photobleaching step itself requires only a visible light source and 10 minutes. Fluorescence scanner experiments characterized visually how effective this method can be. EIS measurements showed a good electrode-to-electrode reproducibility and established the limit of detection of 100 pg/ml of biotin, making this the most sensitive technique investigated during this work.

5 Summary and Outlook

This chapter will briefly summarize the result of this work and give an outlook to each of the topics addressed. Furthermore, the encountered problems will be analyzed and potential solutions and suggestions will be outlined.

5.1 Electrode fabrication procedure

One of the objectives of this work was to develop, implement and optimize the electrode fabrication procedure, thus aiming to achieve a simple, cost-effective, reliable and high-resolution fabrication method. UV photolithography was used with two different mask types, printed masks and dynamic masks created by maskless projection lithography. Three different substrates were investigated throughout this work: plastic, ceramics and glass. Plastic was initially used but as the assays progressively became more complex and time-consuming it became clear that the gold adhesion was insufficient. This led to the used of ceramics (Al_2O_3) which showed a noticeably improved gold adhesion but possessed a very high porosity which increased the nonspecific adsorption considerably. Glass substrates using an adhesion promotion layer consisting of 10 nm of titanium showed an exceptional gold adhesion and very low roughness. Fabrication steps such as spin coating and UV exposure had to be adjusted for each substrate and glass substrates required an additional titanium etching step in order to avoid short circuiting the electrodes. Even though the fabricated electrodes are somewhat large (200 µm gap) it has been shown in the literature that 1 µm line resolution is possible [73]. This feature makes photolithography a suitable fabrication method for the scale-up. Another advantage of this fabrication method is its flexibility and suitability for parallelization.

5.2 Measurement setup

The flow cell fabrication based on casting of PDMS against stereolithographically fabricated epoxy resins has enabled the uncomplicated and fast creation of prototypes for this biosensor platform. The increase of the number of electrodes required analogous changes in the flow cell design. The custom-made measurement electronics was constantly improved throughout this work, increasing the number of channels and the enlarging the operation frequency range. Recent tests performed showed the successful monitoring of 96 channels in parallel. The combination of these expandability features of all the required components of this platform (electrodes, flow cell and measurement electronics) can be used for this electrochemical biosensor platform as well. Such high degree of parallelization is advantageous as many assays (e.g., complimentary DNA library scan for matches) have very large (tens of thousands) sample libraries. The developed platform could be a cheap and flexible alternative to perform these high-throughput assays.

5.3 Surface modification

Surface modification turned out to be the most challenging aspect of this project. Achieving simultaneously all requirements for a very sensitive electrochemical based biosensor was arduous. This layer had to provide functional groups for the binding of the analyte, suppress the nonspecific adsorption while generating only low-impedance layers. These characteristics were only partially fulfilled by the methods found in literature. SAMs did not yield sufficiently low-impedance layers, conductive polymers were not able to suppress the nonspecific adsorption and chemical grafting could not generate reproducible and homogenous layers. The novel photobleaching based surface modification method could fulfill all of these requirements, achieved consistently reproducible results and, furthermore, was the fastest surface modification procedure. The photobleaching step itself only takes 10 minutes. These properties enabled this surface modification strategy to achieve the lowest limit of detection from all the tested methods measured by EIS. By photobleaching fluorescein-biotin and using streptavidin-cy3 as an intermediate molecule this method achieved a limit of detection for biotin of 100 pg/ml (see Figure 64). Further optimizations in order to determine the optimal parameters (electrode design, reagents concentrations, exposure length, wash steps, etc.) for an improved detection limit of this new method still have to be carried out.

5.4 Affinity assay

The affinity assays shown in this work are based on a biotin-streptavidin model system. This system has been chosen as an initial test system due to its known high affinity and robustness. Future experiments should replace this system by a more realistic and application-centered one, i.e., based on antibodies (IgGs) and antigens. The results obtained in this work indicate strongly that this change to more practical systems should be possible with small changes in the described procedure.

6 Conclusion

The objective of this work was to develop an electrochemical biosensor platform which included the electrode fabrication technique, the design of the microfluidic integration and the implementation of a surface modification strategy.
The develop platform consists of a custom made 8-channel measurement electronics which can perform parallel impedance spectrum measurements from 10 to 50.000 Hz on all channels in parallel. The electronics are connected to planar gold electrodes by means of a spring loaded connector slot. In order to improve adhesion gold was sputtered on a glass substrate with a titanium adhesion promoting layer underneath. The electrode structure was created by means of UV photolithography of a positive photoresist (AZ 1512 HS), a gold etch bath in a solution of iodine and potassium iodine and a titanium etch step in piranha solution. The designed microfluidic flow cell has been implemented to house sixteen channels per substrate. Furthermore, it was optimized to hinder channel contamination and direct contact of the tubings with the substrate.
The sensor system was used in measurements which did not require a specific binding of molecules. A chosen application example was the detection of biofilm growth. As a bacterial film colonizes the surface of the electrodes it gradually hinders the current flow between the electrodes, thus leading to a measureable impedance increase. Additionally, it was possible to monitor the biofilm's respiratory activity by means of amperometry. By combining the information obtained from EIS with the data from the amperometric measurements it was possible to implement and test a system which could monitor in real time the growth and respiratory activity of a bacterial colony of *Pseudomonas aeruginosa* [87].

The bare gold electrodes were evaluated for their sensitivity to adsorption of a protein, i.e., BSA. The limit of detection was found to be 1 ng/ml. In order to allow specific binding of an analyte four different surface modification strategies were implemented and evaluated during the course of this work. Due to the simplicity of the method, SAMs were evaluated first. For this five different SAM-forming reagents were tested. The best results were obtained with a short aromatic compound (4-Mercaptobenzoic acid) which was functionalized with streptavidin by means of active ester chemistry and allowed detection of 25 µg/ml of b-BSA using PBS as the buffer. The high layer impedance of this method was responsible for the low sensor sensitivity. The next method evaluated was based on conductive polymers. For this a copolymer of pyrrole and its carboxyl functionalized form (pyrrole-3-carboxylic acid) was electropolymerized onto the gold electrodes. By immobilizing streptavidin to the polymer surface it was possible to detect b-BSA and biotin down to 1 ng/ml and 10 ng/ml, respectively. This method was investigated using not only EIS but also fluorescence microscopy in order to assess layer quality parameters. The disadvantages of this method were the relatively high nonspecific adsorption and the inhomogeneity of the polymer layers. The third method investigated was based on electrochemical grafting of small molecules. By using fluorescence microscopy it was possible to analyze two different surface modification scenarios: Surfaces with high affinity to one specific analyte (positive control) and surfaces with low nonspecific adsorption (negative control). Combining grafting with SAM in order to improve the nonspecific adsorption led to the considerable increase of the layer impedance which rendered this approach unsuitable. These results were confirmed by EIS. Additionally the complexity of the deposition method and the required stringent temperature control led to low reproducibility and the formation of inhomogeneous layers.

The last surface modification method was a novel approach based on photobleaching. By overexposing a molecule labelled with a cheap fluorophore (fluorescein) it was possible to attach this molecule to an unsaturated surface (gold coated with allyl mercaptan). Two different scenarios were created to test this immobilization by means of fluorescence microscopy using streptavidin-cy3 as a label. One surface was functionalized with biotin by photobleaching fluorescein-biotin and a second was functionalized using fluorescein-TRIS. Fluorescence microscopy results clearly indicated the success of this approach. EIS experiments using streptavidin as ligand demonstrated a limit of detection of 100 pg/ml for biotin. Furthermore, high electrode-to-electrode reproducibility was achieved using this surface modification method (see Figure 63).

The developed EIS platform has been successfully applied to affinity binding assays of streptavidin-biotin. The novel surface modification based on photobleaching was evaluated using EIS and achieved a biotin limit of detection of 100 pg/ml when immobilizing streptavidin to the gold electrode's surface. Different analyte specific molecules, e.g., antibodies, coupled to fluorescein can be immobilized to the electrode's surface by means of photobleaching. EIS enables the fast assessment of the electrode's surface electrochemical parameters which correlate directly to a binding event. These properties enable the use of this system as a flexible platform for label-free affinity assays.

A Publication list

2014

L. Pires, N. Braunegger, G. Davidson, C. Neumann and B. E. Rapp
Novel electrochemical biosensor surface modification method based on photobleaching
18th International Conference on Miniaturized Systems for Chemistry and Life Sciences (µTAS), 26-30.10.14, San Antonio, USA, 2014

C. Neumann, E. Wilhelm, T. Duttenhofer, L. Pires, B. E. Rapp
A chemically inert, multichannel Chip-to-World-Interface to connect microfluidic chips
SPIE Photonics West, 01.-06.02.2014, San Francisco, USA, 2014.

T.M. Nargang, L. Brockmann, P. Nikolov, D. Schild, D. Helmer, N. Keller, K. Sachsenheimer, E. Wilhelm, L. Pires, M. Dirschka, A. Kolew, M. Schneider, M. Worgull, S. Giselbrecht, C. Neumann, B.E. Rapp.
Liquid polystyrene: a room-temperature photocurable soft lithography compatible pour-and-cure-type polystyrene
Lab Chip, DOI: 10.1039/C4LC00045E, 2014.

2013

C. Neumann, A. Voigt, L. Pires, B. E. Rapp
Design and characterization of a platform for thermal actuation of up to 588 microfluidic valves
Microfluidics and Nanofluidics, 14(1-2), 177-186, 2013.

L. Pires, K. Sachsenheimer, T. Kleintschek, A. Waldbaur, T. Schwartz, B. E. Rapp
Online Monitoring of Biofilm Growth and Activity Using a Combined Multi-channel Impedimetric and Amperometric Sensor
Biosensors and Bioelectronics, 47, 157–163, 2013.

E. Wilhelm, C. Neumann, T. Duttenhofer, L. Pires, B. E. Rapp
Connecting microfluidic chips using a chemically inert, reversible, multichannel Chip-to-World-Interface
Lab Chip, 13, 4343-4351, 2013.

B. E. Rapp, L. Carneiro, A. Voigt
Device for controlling the flow of fluids through microfluidic channels US patent application number US 8480974 B2 (also published as US 20110023971 A1), 2013.

B. E. Rapp, L. Carneiro, A. Voigt
Vorrichtung zur Erzeugung einer mikrofluidischen Kanalstruktur in einer Kammer, Verfahren zu ihrer Betrieb und ihre Verwendung
European patent number EP 2279788 A3, 2013 (in revision).

B. E. Rapp, L. Carneiro, A. Voigt

Vorrichtung zum Steuern des Durchflusses von Fluiden durch mikrofluidische Kanäle, Verfahren zu ihrem Betrieb und ihrer Verwendung
European patent number EP 2279789 A3, 2013 (in revision).

2012

L. Pires, A. Heckel, K. Sachsenheimer, B.E. Rapp
Multichannel Impedimetric Biosensor Platform for Label-Free Affinity Assays Using Electrically Conductive Functional Polymers
16th International Conference on Miniaturized Systems for Chemistry and Life Sciences (µTAS), 28.10-01.11.2012, Okinawa, Japan, 2012.

2011

K. Sachsenheimer, L. Pires, M. Adamek, T. Schwartz, B. E. Rapp
Monitoring biofilm growth using a scalable multichannel impedimetric biosensor
15th International Conference on Miniaturized Systems for Chemistry and Life Sciences (µTAS), 02.-06.10.2011, Seattle, USA, 2011.

B. E. Rapp, L. Carneiro, A. Voigt
Vorrichtung zur Erzeugung einer mikrofluidischen Kanalstruktur in einer Kammer, Verfahren zu ihrer Betrieb und ihre Verwendung
German patent number DE 102009035291 B4, 2011.

B. E. Rapp, L. Carneiro, A. Voigt
Vorrichtung zum Steuern des Durchflusses von Fluiden durch mikrofluidische Kanäle, Verfahren zu ihrem Betrieb und ihrer Verwendung
German patent number DE 102009035292 A1, 2011 (in revision).

B. E. Rapp, L. Carneiro, A. Voigt
Device for creating a microfluidic channel structure in a chamber,
US patent application number US 20110023970 A1, 2011 (in revision).

2009

B. E. Rapp, L. Carneiro, K. Länge, M. Rapp
An indirect microfluidic flow injection analysis (FIA) system allowing diffusion free pumping of liquids by using tetradecane as intermediate liquid
Lab Chip, 9, 354-6, 2009.

2007

B. E. Rapp, L. Carneiro, K. Länge, M. Rapp
Vorrichtung und Verfahren zur Zuführung eines Flüssigkeitsstroms aus mindestens zwei Flüssigkeitsabschnitten in eine Messzelle
German patent number DE 102007032951 B4, 2007.

B Curriculum vitae

Personal data

Name	Leonardo Pires Carneiro
Address	Badener Str., 3 76227, Karlsruhe
Birth date	26.11.1984
Place of birth	Belo Horizonte, Brazil
Nationality	Brazilian

Education

1992-1995	Primary school Colégio Pitagoras - Cidade Jardim, Belo Horizonte, Brazil – Overall grade: 1.7
1995-2002	Secondary school Colégio Pitagoras - Cidade Jardim, Belo Horizonte, Brazil – Overall grade: 1.5
01/2003 – 07/2008	Bachelor in Electrical Engineering by the Federal University of Minas Gerais (UFMG), Minas Gerais, Brazil. Overall grade 1.9
10/2006 – 08/2007	Academic exchange program at TU-Karlsruhe

10/2008 – 11/2010	Master of Science in Electrical Engineering and Information Technology at Karlsruhe Institute of Technology (KIT). Master thesis entitled: "Feasibility Study of a 122 GHz Antenna-in-Package in LTCC Technology" Grade 1.0 Overall grade: 1.5
01/2011-03/2014	PhD at the Institute of Microstructure Technology, KIT: "Development of an Electrochemical Biosensor Platform and a Suitable Low-Impedance Surface Modification Strategy" Date of defense: 19.08.2014
09/2012 – 12/2012	Academic exchange program at Texas A&M University, College Station, TX, USA

C Bibliography

1. Turner, A.P.F., *Biosensors: sense and sensibility.* Chemical Society Reviews, 2013. **42**(8): p. 3184-3196.
2. Guilbault, G.G., et al., *Biosensors - 42 years and counting.* Analytical Letters, 2004. **37**(8): p. 1481-1496.
3. *Global Health Expenditure Database.* [Webpage] [cited 2013 16.12]; Available from: http://apps.who.int/nha/database.
4. Klonoff, D.C., *Continuous glucose monitoring roadmap for 21st century diabetes therapy.* Diabetes care, 2005. **28**(5): p. 1231-1239.
5. Sang, S., W. Zhang, and Y. Zhao, *Review on the Design Art of Biosensors.* State of the Art in Biosensors - General Aspects. 2013.
6. Luong, J.H.T., K.B. Male, and J.D. Glennon, *Biosensor technology: Technology push versus market pull.* Biotechnology Advances, 2008. **26**(5): p. 492-500.
7. Pancrazio, J.J., et al., *Development and Application of Cell-Based Biosensors.* Annals of Biomedical Engineering, 1999. **27**(6): p. 697-711.
8. Luppa, P.B., L.J. Sokoll, and D.W. Chan, *Immunosensors—principles and applications to clinical chemistry.* Clinica Chimica Acta, 2001. **314**(1–2): p. 1-26.
9. Wang, J., *Glucose Biosensors: 40 Years of Advances and Challenges.* Electroanalysis, 2001. **13**(12): p. 983-988.
10. Wang, J., *From DNA biosensors to gene chips.* Nucleic Acids Research, 2000. **28**(16): p. 3011-3016.
11. Clark, L.C., *MONITOR AND CONTROL OF BLOOD AND TISSUE OXYGEN TENSIONS.* Transactions American Society for Artificial Internal Organs, 1956. **2**: p. 41-&.
12. Clark, L.C. and C. Lyons, *ELECTRODE SYSTEMS FOR CONTINUOUS MONITORING IN CARDIOVASCULAR SURGERY.* Annals of the New York Academy of Sciences, 1962. **102**(1): p. 29-45.

13. Guilbault, G.G. and J.G. Montalvo, *Urea-specific enzyme electrode.* Journal of the American Chemical Society, 1969. **91**(8): p. 2164-2165.
14. Cooney, C.L., et al., *The Thermal Enzyme Probe — A Novel Approach to Chemical Analysis*, in *Enzyme Engineering Volume 2*, E.K. Pye and L. Wingard, Jr., Editors. 1974, Springer US. p. 411-417.
15. Lubbers, D.W. and N. Opitz, *GE))PCO2-OPTODE-PO2-OPTODE - NEW PROBE FOR MEASUREMENT OF PCO2 OR PO2 IN FLUIDS AND GASES.* Zeitschrift Fur Naturforschung C-a Journal of Biosciences, 1975. **30**(7-8): p. 532-533.
16. Divies, C., *REMARKS ON ETHANOL OXIDATION BY AN ACETOBACTER-XYLINUM MICROBIAL ELECTRODE.* Annales De Microbiologie, 1975. **A126**(2): p. 175-186.
17. Kricka, L. and J. Savory, *A Guide to the History of Clinical Chemistry.* Clinical Chemistry, 2011.
18. Lequin, R.M., *Enzyme immunoassay (EIA)/enzyme-linked immunosorbent assay (ELISA).* Clinical Chemistry, 2005. **51**(12): p. 2415-2418.
19. Cass, A.E.G., et al., *Ferrocene-mediated enzyme electrode for amperometric determination of glucose.* Analytical Chemistry, 1984. **56**(4): p. 667-671.
20. Newman, J. and S. Setford, *Enzymatic biosensors.* Molecular Biotechnology, 2006. **32**(3): p. 249-268.
21. Liedberg, B., C. Nylander, and I. Lundstrom, *SURFACE-PLASMON RESONANCE FOR GAS-DETECTION AND BIOSENSING.* Sensors and Actuators, 1983. **4**(2): p. 299-304.
22. Liedberg, B., C. Nylander, and I. Lunström, *Surface plasmon resonance for gas detection and biosensing.* Sensors and Actuators, 1983. **4**(0): p. 299-304.
23. Bakker, E. and E. Pretsch, *Potentiometric sensors for trace-level analysis.* Trac-Trends in Analytical Chemistry, 2005. **24**(3): p. 199-207.
24. Borgmann, S., et al., *Amperometric Biosensors*, in *Advances in Electrochemical Science and Engineering.* 2011, Wiley-VCH Verlag GmbH & Co. KGaA. p. 1-83.

25. Lisdat, F. and D. Schäfer, *The use of electrochemical impedance spectroscopy for biosensing.* Analytical and Bioanalytical Chemistry, 2008. **391**(5): p. 1555-1567.
26. Macdonald, D.D., *Reflections on the history of electrochemical impedance spectroscopy.* Electrochimica Acta, 2006. **51**(8–9): p. 1376-1388.
27. Bănică, F.-G., *Dynamic Electrochemistry Transduction Methods*, in *Chemical Sensors and Biosensors.* 2012, John Wiley & Sons, Ltd. p. 258-313.
28. Randles, J.E.B., *Kinetics of rapid electrode reactions.* Discussions of the Faraday Society, 1947. **1**(0): p. 11-19.
29. Grieshaber, D., et al., *Electrochemical Biosensors - Sensor Principles and Architectures.* Sensors, 2008. **8**(3): p. 1400-1458.
30. Lucklum, R. and P. Hauptmann, *Acoustic microsensors-the challenge behind microgravimetry.* Analytical and Bioanalytical Chemistry, 2006. **384**(3): p. 667-682.
31. Daniels, J.S. and N. Pourmand, *Label-Free Impedance Biosensors: Opportunities and Challenges.* Electroanalysis, 2007. **19**(12): p. 1239-1257.
32. Gauglitz, G. and G. Proll, *Strategies for label-free optical detection*, in *Biosensing for the 21st Century*, R. Renneberg and F. Lisdat, Editors. 2008. p. 395-432.
33. Ramanathan, K. and B. Danielsson, *Principles and applications of thermal biosensors.* Biosensors & Bioelectronics, 2001. **16**(6): p. 417-423.
34. Peluso, P., et al., *Optimizing antibody immobilization strategies for the construction of protein microarrays.* Analytical Biochemistry, 2003. **312**(2): p. 113-124.
35. González, M., et al., *Interaction of Biotin with Streptavidin: THERMOSTABILITY AND CONFORMATIONAL CHANGES UPON BINDING.* Journal of Biological Chemistry, 1997. **272**(17): p. 11288-11294.
36. Ceri, H., et al., *The Calgary Biofilm Device: New technology for rapid determination of antibiotic susceptibilities of bacterial biofilms.* Journal of Clinical Microbiology, 1999. **37**(6): p. 1771-1776.

37. Costerton, J.W., P.S. Stewart, and E.P. Greenberg, *Bacterial biofilms: A common cause of persistent infections.* Science, 1999. **284**(5418): p. 1318-1322.
38. Tyler, P.A. and K.C. Marshall, *Microbial oxidation of manganese in hydro-electric pipelines.* Antonie van Leeuwenhoek, 1967. **33**(1): p. 171-183.
39. Flemming, H.-C., et al., *Biofouling—the Achilles heel of membrane processes.* Desalination, 1997. **113**(2): p. 215-225.
40. Bott, T.R., *Fouling of heat exchangers*. 1995: Access Online via Elsevier.
41. Cooksey, K.E. and B. Wigglesworthcooksey, *ADHESION OF BACTERIA AND DIATOMS TO SURFACES IN THE SEA - A REVIEW.* Aquatic Microbial Ecology, 1995. **9**(1): p. 87-96.
42. Reipa, V., J. Almeida, and K.D. Cole, *Long-term monitoring of biofilm growth and disinfection using a quartz crystal microbalance and reflectance measurements.* Journal of Microbiological Methods, 2006. **66**(3): p. 449-459.
43. Jenkins, A.T.A., et al., *Study of the attachment of Pseudomonas aeruginosa on gold and modified gold surfaces using surface plasmon resonance.* Biotechnology Progress, 2004. **20**(4): p. 1233-1236.
44. Young Wook, K., et al. *A bacterial biofilm Surface Acoustic Wave sensor for real time biofilm growth monitoring.* in *Sensors, 2010 IEEE*. 2010.
45. Chandra, J., et al., *Biofilm formation by the fungal pathogen Candida albicans: Development, architecture, and drug resistance.* Journal of Bacteriology, 2001. **183**(18): p. 5385-5394.
46. Beyer, M., et al., *A novel glass slide-based peptide array support with high functionality resisting non-specific protein adsorption.* Biomaterials, 2006. **27**(18): p. 3505-3514.
47. Lasseter, T.L., et al., *Covalently Modified Silicon and Diamond Surfaces: Resistance to Nonspecific Protein Adsorption and Optimization for Biosensing.* Journal of the American Chemical Society, 2004. **126**(33): p. 10220-10221.
48. Desai, N.P., S.F.A. Hossainy, and J.A. Hubbell, *Surface-immobilized polyethylene oxide for bacterial repellence.* Biomaterials, 1992. **13**(7): p. 417-420.

49. Yoshioka, K., et al., *Suppression of non-specific adsorption using densified tri (ethylene glycol) alkanethiols: monolayer characteristics evaluated by electrochemical measurements.* Analytical sciences: the international journal of the Japan Society for Analytical Chemistry, 2010. **26**(1): p. 33-37.
50. Wink, T., et al., *Self-assembled Monolayers for Biosensors.* Analyst, 1997. **122**(4): p. 43R-50R.
51. Cosnier, S. and M. Holzinger, *Electrosynthesized polymers for biosensing.* Chemical Society Reviews, 2011. **40**(5): p. 2146-2156.
52. Vidal, J.C., E. Garcia, and J.R. Castillo, *In situ preparation of a cholesterol biosensor: entrapment of cholesterol oxidase in an overoxidized polypyrrole film electrodeposited in a flow system - Determination of total cholesterol in serum.* Analytica Chimica Acta, 1999. **385**(1-3): p. 213-222.
53. Mahouche-Chergui, S., et al., *Aryl diazonium salts: a new class of coupling agents for bonding polymers, biomacromolecules and nanoparticles to surfaces.* Chemical Society Reviews, 2011. **40**(7): p. 4143-4166.
54. Ouerghi, O., et al., *Impedimetric immunosensor using avidin–biotin for antibody immobilization.* Bioelectrochemistry, 2002. **56**(1–2): p. 131-133.
55. Lai, S., et al., *Design of a compact disk-like microfluidic platform for enzyme-linked immunosorbent assay.* Analytical chemistry, 2004. **76**(7): p. 1832-1837.
56. Holden, M.A. and P.S. Cremer, *Light activated patterning of dye-labeled molecules on surfaces.* Journal of the American Chemical Society, 2003. **125**(27): p. 8074-8075.
57. Scrimgeour, J., et al., *Photobleaching-activated micropatterning on self-assembled monolayers.* Journal of Physics: Condensed Matter, 2010. **22**(19): p. 194103.
58. Waldbaur, A., et al., *Maskless projection lithography for the fast and flexible generation of grayscale protein patterns.* Small, 2012. **8**(10): p. 1570-1578.
59. Grabarek, Z. and J. Gergely, *Zero-length crosslinking procedure with the use of active esters.* Analytical Biochemistry, 1990. **185**(1): p. 131-135.

60. Shirakawa, H., et al., *SYNTHESIS OF ELECTRICALLY CONDUCTING ORGANIC POLYMERS - HALOGEN DERIVATIVES OF POLYACETYLENE, (CH)X.* Journal of the Chemical Society-Chemical Communications, 1977(16): p. 578-580.
61. Ates, M., *Review study of electrochemical impedance spectroscopy and equivalent electrical circuits of conducting polymers on carbon surfaces.* Progress in Organic Coatings, 2011. **71**(1): p. 1-10.
62. Lorenz, H., et al., *SU-8: a low-cost negative resist for MEMS.* Journal of Micromechanics and Microengineering, 1997. **7**(3): p. 121.
63. *General Properties of AZ®/ TI Photoresists.* [Webpage] 2007 [cited 2013 21.08]; Available from: http://www.microchemicals.eu/technical_information/photoresist_properties.pdf.
64. Dill, F.H., et al., *Characterization of positive photoresist.* Electron Devices, IEEE Transactions on, 1975. **22**(7): p. 445-452.
65. Wolff, L. and R. Krüche, *Über Diazoanhydride (1,2,3-Oxydiazole oder Diazoxyde) und Diazoketone.* Justus Liebigs Annalen der Chemie, 1912. **394**(1): p. 23-59.
66. *Exposure of photoresists.* 2010 [cited 2013 29.08]; Available from: http://www.microchemicals.eu/technical_information/exposure_photoresist.pdf.
67. Sigmund, P., *Theory of Sputtering. I. Sputtering Yield of Amorphous and Polycrystalline Targets.* Physical Review, 1969. **184**(2): p. 383-416.
68. Lichtman, J.W. and J.-A. Conchello, *Fluorescence microscopy.* Nat Meth, 2005. **2**(12): p. 910-919.
69. Waldbaur, A., et al., *Let there be chip-towards rapid prototyping of microfluidic devices: one-step manufacturing processes.* Analytical Methods, 2011. **3**(12): p. 2681-2716.
70. Xia, Y.N. and G.M. Whitesides, *Soft lithography.* Annual Review of Materials Science, 1998. **28**: p. 153-184.
71. Loeb, G.E., et al., *HISTOLOGICAL REACTION TO VARIOUS CONDUCTIVE AND DIELECTRIC FILMS CHRONICALLY IMPLANTED IN SUBDURAL SPACE.* Journal of Biomedical Materials Research, 1977. **11**(2): p. 195-210.

72. Chang, T.Y., et al., *Cell and protein compatibility of parylene-C surfaces.* Langmuir, 2007. **23**(23): p. 11718-11725.
73. Waldbaur, A., *Entwicklung eines maskenlosen Fotolithographiesystems zum Einsatz im Rapid Prototyping in der Mikrofluidik und zur gezielten Oberflächenfunktionalisierung*. 2013: KIT Scientific Publishing.
74. Serway, R.A. and J.W. Jewett, *Serway's principles of physics: a calculus-based text*. Vol. 1. 2006: Cengage Learning.
75. Poon, V.K.M. and A. Burd, *In vitro cytotoxity of silver: implication for clinical wound care.* Burns, 2004. **30**(2): p. 140-147.
76. Wataha, J.C., P.E. Lockwood, and A. Schedle, *Effect of silver, copper, mercury, and nickel ions on cellular proliferation during extended, low-dose exposures.* Journal of Biomedical Materials Research, 2000. **52**(2): p. 360-364.
77. Connor, E.E., et al., *Gold nanoparticles are taken up by human cells but do not cause acute cytotoxicity.* Small, 2005. **1**(3): p. 325-327.
78. Atkins, P. and J. De Paula, *Elements of physical chemistry*. 2012: Oxford University Press.
79. Bagchi, D., et al., *Cytotoxicity and oxidative mechanisms of different forms of chromium.* Toxicology, 2002. **180**(1): p. 5-22.
80. Tabeling, P., *Introduction to microfluidics*. Repr. ed. 2006, Oxford [u.a.]: Oxford University Press. VII, 301 S.
81. *UV-Lampe_SL400_Muttergerät_041109*. [Dataset] 2009 [cited 2013 04.09].
82. Williams, K.R., K. Gupta, and M. Wasilik, *Etch rates for micromachining processing - Part II.* Journal of Microelectromechanical Systems, 2003. **12**(6): p. 761-778.
83. *Gold Etching*. [Webpage] 2013 [cited 2013 20.11]; Available from: http://www.microchemicals.com/technical_information/gold_etching.pdf.
84. Singh, B. and N.A. Surplice, *The electrical resistivity and resistance-temperature characteristics of thin titanium films.* Thin Solid Films, 1972. **10**(2): p. 243-253.
85. Williams, K.R. and R.S. Muller, *Etch rates for micromachining processing.* Microelectromechanical Systems, Journal of, 1996. **5**(4): p. 256-269.

86. Richardson, D.J., *Bacterial respiration: a flexible process for a changing environment.* Microbiology, 2000. **146**(3): p. 551-571.
87. Pires, L., et al., *Online monitoring of biofilm growth and activity using a combined multi-channel impedimetric and amperometric sensor.* Biosensors and Bioelectronics, 2013. **47**(0): p. 157-163.
88. Fischer, L.M., et al., *Gold cleaning methods for electrochemical detection applications.* Microelectronic Engineering, 2009. **86**(4–6): p. 1282-1285.
89. Li, D., *Encyclopedia of microfluidics and nanofluidics*. Vol. 1. 2008: Springer.
90. SerenaáChiriac, M. and R. Elenaálonescu, *EIS microfluidic chips for flow immunoassay and ultrasensitive cholera toxin detection.* Lab on a Chip, 2011. **11**(4): p. 658-663.
91. Montrose, A., et al., *Impedimetric immunosensor for the detection of circulating pro-inflammatory monocytes as infection markers.* Biosensors and Bioelectronics, 2013. **49**: p. 305-311.
92. Steude, A., et al., *An electrode array for electrochemical immuno-sensing using the example of impedimetric tenascin C detection.* Lab on a Chip, 2011. **11**(17): p. 2884-2892.
93. Chebil, S., et al., *Electrochemical detection of d-dimer as deep vein thrombosis marker using single-chain d-dimer antibody immobilized on functionalized polypyrrole.* Biosensors and Bioelectronics, 2010. **26**(2): p. 736-742.
94. Chiriaco, M.S., et al., *EIS microfluidic chips for flow immunoassay and ultrasensitive cholera toxin detection.* Lab on a Chip, 2011. **11**(4): p. 658-663.
95. Chen, C.-S., et al., *Development of a label-free impedance biosensor for detection of antibody–antigen interactions based on a novel conductive linker.* Biosensors and Bioelectronics, 2011. **26**(6): p. 3072-3076.
96. Truong, L.T., et al., *Labelless impedance immunosensor based on polypyrrole–pyrolecarboxylic acid copolymer for hCG detection.* Talanta, 2011. **85**(5): p. 2576-2580.
97. Lee, S.-W. and P.E. Laibinis, *Protein-resistant coatings for glass and metal oxide surfaces derived from oligo(ethylene glycol)-terminated alkyltrichlorosilanes.* Biomaterials, 1998. **19**(18): p. 1669-1675.

Schriften des Instituts für Mikrostrukturtechnik
am Karlsruher Institut für Technologie (KIT)

ISSN 1869-5183

Herausgeber: Institut für Mikrostrukturtechnik

Die Bände sind unter www.ksp.kit.edu als PDF frei verfügbar
oder als Druckausgabe zu bestellen.

Band 1 Georg Obermaier
Research-to-Business Beziehungen: Technologietransfer durch Kommunikation von Werten (Barrieren, Erfolgsfaktoren und Strategien). 2009
ISBN 978-3-86644-448-5

Band 2 Thomas Grund
Entwicklung von Kunststoff-Mikroventilen im Batch-Verfahren. 2010
ISBN 978-3-86644-496-6

Band 3 Sven Schüle
Modular adaptive mikrooptische Systeme in Kombination mit Mikroaktoren. 2010
ISBN 978-3-86644-529-1

Band 4 Markus Simon
Röntgenlinsen mit großer Apertur. 2010
ISBN 978-3-86644-530-7

Band 5 K. Phillip Schierjott
Miniaturisierte Kapillarelektrophorese zur kontinuierlichen Überwachung von Kationen und Anionen in Prozessströmen. 2010
ISBN 978-3-86644-523-9

Band 6 Stephanie Kißling
Chemische und elektrochemische Methoden zur Oberflächenbearbeitung von galvanogeformten Nickel-Mikrostrukturen. 2010
ISBN 978-3-86644-548-2

Schriften des Instituts für Mikrostrukturtechnik
am Karlsruher Institut für Technologie (KIT)

ISSN 1869-5183

Band 7 Friederike J. Gruhl
Oberflächenmodifikation von Surface Acoustic Wave (SAW) Biosensoren für biomedizinische Anwendungen. 2010
ISBN 978-3-86644-543-7

Band 8 Laura Zimmermann
Dreidimensional nanostrukturierte und superhydrophobe mikrofluidische Systeme zur Tröpfchengenerierung und -handhabung. 2011
ISBN 978-3-86644-634-2

Band 9 Martina Reinhardt
Funktionalisierte, polymere Mikrostrukturen für die dreidimensionale Zellkultur. 2011
ISBN 978-3-86644-616-8

Band 10 Mauno Schelb
Integrierte Sensoren mit photonischen Kristallen auf Polymerbasis. 2012
ISBN 978-3-86644-813-1

Band 11 Daniel Auernhammer
Integrierte Lagesensorik für ein adaptives mikrooptisches Ablenksystem. 2012
ISBN 978-3-86644-829-2

Band 12 Nils Z. Danckwardt
Pumpfreier Magnetpartikeltransport in einem Mikroreaktionssystem: Konzeption, Simulation und Machbarkeitsnachweis. 2012
ISBN 978-3-86644-846-9

Band 13 Alexander Kolew
Heißprägen von Verbundfolien für mikrofluidische Anwendungen. 2012
ISBN 978-3-86644-888-9

Schriften des Instituts für Mikrostrukturtechnik
am Karlsruher Institut für Technologie (KIT)

ISSN 1869-5183

Band 14 Marko Brammer
Modulare Optoelektronische Mikrofluidische Backplane. 2012
ISBN 978-3-86644-920-6

Band 15 Christiane Neumann
Entwicklung einer Plattform zur individuellen Ansteuerung von Mikroventilen und Aktoren auf der Grundlage eines Phasenüberganges zum Einsatz in der Mikrofluidik. 2013
ISBN 978-3-86644-975-6

Band 16 Julian Hartbaum
Magnetisches Nanoaktorsystem. 2013
ISBN 978-3-86644-981-7

Band 17 Johannes Kenntner
Herstellung von Gitterstrukturen mit Aspektverhältnis 100 für die Phasenkontrastbildgebung in einem Talbot-Interferometer. 2013
ISBN 978-3-7315-0016-2

Band 18 Kristina Kreppenhofer
Modular Biomicrofluidics - Mikrofluidikchips im Baukastensystem für Anwendungen aus der Zellbiologie. 2013
ISBN 978-3-7315-0036-0

Band 19 Ansgar Waldbaur
Entwicklung eines maskenlosen Fotolithographiesystems zum Einsatz im Rapid Prototyping in der Mikrofluidik und zur gezielten Oberflächenfunktionalisierung. 2013
ISBN 978-3-7315-0119-0

Band 20 Christof Megnin
Formgedächtnis-Mikroventile für eine fluidische Plattform. 2013
ISBN 978-3-7315-0121-3

Schriften des Instituts für Mikrostrukturtechnik
am Karlsruher Institut für Technologie (KIT)

ISSN 1869-5183

Band 21 Srinivasa Reddy Yeduru
Development of Microactuators Based on the Magnetic Shape Memory Effect. 2013
ISBN 978-3-7315-0125-1

Band 22 Michael Röhrig
Fabrication and Analysis of Bio-Inspired Smart Surfaces. 2014
ISBN 978-3-7315-0163-3

Band 23 Taleieh Rajabi
Entwicklung eines mikrofluidischen Zweikammer-Chipsystems mit integrierter Sensorik für die Anwendung in der Tumorforschung. 2014
ISBN 978-3-7315-0220-3

Band 24 Frieder Märkle
Laserbasierte Verfahren zur Herstellung hochdichter Peptidarrays. 2014
ISBN 978-3-7315-0222-7

Band 25 Tobias Meier
Magnetoresistive and Thermoresistive Scanning Probe Microscopy with Applications in Micro- and Nanotechnology. 2014
ISBN 978-3-7315-0253-1

Band 26 Felix Marschall
Entwicklung eines Röntgenmikroskops für Photonenenergien von 15 keV bis 30 keV. 2014
ISBN 978-3-7315-0263-0

Band 27 Leonardo Pires Carneiro
Development of an Electrochemical Biosensor Platform and a Suitable Low-Impedance Surface Modification Strategy. 2014
ISBN 978-3-7315-0272-2